PRAISE FOR *HIGH-TECH*

"Pete Dulcamara is a true innovator, and he presents a remarkable blend of wisdom and insights drawn from his decades of leadership in research and development and as a global leader in business, technology, and innovation. This calls on the next generation to embrace science, engineering, and entrepreneurship to drive meaningful progress and it inspires young minds to build sustainable businesses that serve both people and planet."

—Andrew N. Liveris, Former Chairman & CEO, Dow Chemical, President Brisbane 2032 Olympic and Paralympic Games Organising Committee

"*High-Tech Heroes* is an essential and inspiring call to action for *Gen Z* and future leaders, demonstrating how AI, robotics, and circular economies can drive sustainable innovation and global change. Pete Dulcamara redefines success, urging readers to build and nurture innovation as a continuous capability and not just a one-time event, leveraging technology for humanity-centric solutions that balance purpose, profit, and prosperity."

—Cheryl Perkins, Founder & CEO, Innovationedge, LLC

"Real change happens when we break down silos. *High-Tech Heroes* shows how the next generation can bridge the gap between corporations and impact-driven startups to create solutions that truly matter."

—James Sancto, Co-Founder & CEO, We Make Change

"Innovation doesn't happen without the right talent. *High-Tech Heroes* underscores the vital connection between talent and innovation. A compelling guide for building teams that drive transformative change. It reflects the leadership qualities needed to nurture talent and deliver breakthrough results in today's dynamic business environment."

—Tony Palmer, Former President, Global Brands & Innovation, Kimberly-Clark

"To make an impact, you have to challenge assumptions, embrace change, and get unstuck. In *High-Tech Heroes*, Pete pushes young leaders to challenge the status quo. It provides the tools and mindset needed to drive innovation that matters."

—Clive Sirkin, Host of UNSTUCK! Podcast / Board Member / Investor / Strategic Advisor

"SDG #5—Gender Equality—isn't just another development goal; it's a non-negotiable target for a thriving world. We all have a role in leveling the playing field so women, girls, and future generations can step into the race equipped with the knowledge, tools, technology, leadership, and ethics needed to drive real change. This is a call to action for young leaders: don't lose faith. Innovate, challenge the status quo, and create solutions that keep women and girls in the race—because opportunity should be based on merit, not gender. Equality isn't optional—it's the foundation for a better future for everyone."

—Diana Sierra, Founder & CEO, Be Girl INC | Be Girl Mozambique LTD

"For billions of years, nature has perfected the art of exponential abundance—it's time we learn from her. *High-Tech Heroes* dares the next generation to harness science and the unprecedented power of AI to reimagine business, not by "exploiting" our planet's resources, but by working in harmony with its limitless, symbiotic gifts. A transformative read for those determined to shape a truly sustainable future."

—Serge Rogasik | President and CEO, Kensing Solutions LLC

"A visionary guide to sustainable innovation that highlights how young leaders can shape a more sustainable future through materials science and cutting-edge innovation. It encourages the next generation to leverage science for environmental progress while building successful businesses."

—James Gibson, CEO, VOID Technologies

"The true power of this generation lies not in the AI technology it wields, but in the courage to wield it with purpose. *High-Tech Heroes* is a manifesto for those who dare to redefine success. It empowers *Gen Z* to harness the transformative power of AI to solve the world's toughest challenges, from climate change to sustainable growth. It's a blueprint for building a future where intelligence—both human and artificial—works in harmony to create opportunity, equity, and a legacy worth leaving. This is not just a book; it's a battle cry for those who refuse to inherit a broken world and choose to take hold of the future."

—Israel Squires, CEO & Co-Founder, Midpoint Consulting

"AI is redefining the future of industry, and sustained innovation is key to long-term success. *High-Tech Heroes* is an essential guide for the next generation of innovators, showing them how to not just embrace new technologies, but to build the ecosystems that make transformation lasting."

—Sahitya Senapathy, CEO, Endeavor.ai

"Purpose-driven companies are the future, and young innovators have the opportunity to build them. *High-Tech Heroes* highlights how startups can merge purpose with profit using 21st-century business models. It encourages the next generation to create impactful ventures that address global challenges while thriving economically."

—Nitin Parekh, Founding Director, High Impact Technology (HIT) Fund | Director, Sustainability Strategic Alliances, Stanford University

"The future belongs to those who think exponentially. *High-Tech Heroes* equips the next generation with the mindset and strategies to propel human enterprise and humanity forward."

—Nayan Shah, Co-Founder & CEO, FutureBridge

HIGH-TECH HEROES

ISBN Paperback: 978-1-963271-78-2
ISBN Ebook: 978-1-963271-79-9

Published by Armin Lear Press, Inc.
215 W Riverside Drive, #4362
Estes Park, CO 80517

HIGH-TECH HEROES

WHY GEN Z IS OUR LAST AND BEST CHANCE TO SAVE THE PLANET

PETE DULCAMARA

For Gina, Rachel, and Peter, together forever.

CONTENTS

ACKNOWLEDGMENTS

This book defines humanity-centric innovation and lays out a path for implementing it. My first acknowledgment is to you, dear reader, with the hope that it inspires you to take meaningful action toward solving the world's biggest problems in an economically viable way.

I am deeply grateful to my children, Rachel and Peter, who have been the impetus for my personal mission to create businesses that improve people's lives. Their presence in my life has given me a purpose and a vision that shaped the heart of this book. To my wife, Gina, my endless gratitude for her ever-present love and support. This journey would not have been possible without her encouragement and belief in my work.

I also want to recognize Jon Mathews, Clive Sirkin, and Cheryl Perkins for their ongoing encouragement and accountability, never failing to ask, "How's the book coming?" Their check-ins were both grounding and motivating, pushing me forward when I needed it most.

Finally, my sincere thanks to Ruby Peru and Maryann Karinch, who took my vision, thoughts, and ideas and transformed them into words on the page and into the book in your hands. Their skill and dedication brought this book to life in a way that I couldn't have achieved alone.

INTRODUCTION

MY PLEA TO GENERATION Z

Nothing in life is to be feared, it is only to be understood. Now is the time to understand more, so that we may fear less.

— Marie Curie, physicist and chemist whose pioneering research on radioactivity earned her two Nobel Prizes—the first person to achieve this honor

I'm old enough to remember, as a child, watching Neil Armstrong land on the moon and marveling at what the world's greatest innovation could accomplish. It was a pivotal moment for me and may indeed have inspired my lifelong passion to improve peoples' lives through science and technology. Back then, sending a man to the moon seemed like a wild flight of fancy, as if a thousand brilliant engineers had asked one another, "What if? What if?" so many times that they came up with the most extraordinary "What if?" of all and then, miraculously, achieved it.

I wanted to be one of those amazing people, one day—to change the world with extraordinary innovation. That may sound a bit old-school. You've held advanced technology in your hands your entire life—not "advanced" to you, but to people born before 1995—and when it comes to the space race, I'll bet even private, manned space flight barely impresses you. In fact, the outstanding technological "What ifs?" of your generation weren't even televised, even though they changed society dramatically . . . because they've primarily been in the medical field.

When it comes to advances in medical technology, you'll certainly never forget the rapidly developed COVID-19 mRNA vaccine that saved innumerable lives after many horrifying months of the pandemic. In fact, Gen Z has become so used to the rapid progress of technology that no matter what incredible strides are made, you're seldom overwhelmed and impressed the way I was by the moon landing. Still, for those of you eager to succeed in tech-oriented fields, it must be inspiring to think of joining the revolution to build a post-globalist society, end environmental destruction, and build robots that will change everything about how the world works by the time you're a grandparent.

Even though I'm a Boomer, which is a generation born after World War II and until the mid-1960s, my unusual life experience has caused me to deeply relate to Gen Z values. People say that, once upon a time, Boomers changed society, started it down a whole new path.

Gen Zers will do the same, but in a far more radical way. I stand by that belief and logical conclusion, both of which drive the message of this book.

If you hope for a life that blends personal fulfillment with doing good for the planet, we're not dissimilar. Like many in both of our generations, I have always wanted to work toward social equity, but who would have known a degree in chemical engineering would lead me to strive for women's empowerment? I have achieved this complex set of goals by allowing the work I do to take me in unexpected directions, going with the flow and sometimes against the grain. Yet, I have also mastered life's direction—as I've noticed so many Gen Zers aspire to do—rejecting work that didn't fulfill personal ambitions. That's why (setting the issue of technology aside) your generation and mine have a lot in common.

A Lot Rides on Your Shoulders

Like most of you, I enjoy technology and believe in its value. I use it for fun and work, and I know what you know: today's radically different technology can solve a lot of global problems, create stability for people that desperately need it, and make it possible for your generation to get paid to change the world. *Or* it can build a world of impersonal robots that mine precious minerals from a troubled Earth and intensify global warming toward a smog-choked apocalypse.

Gen Z will decide which way it goes.

When you're at the edge of that cliff we call adolescence, setting your life's trajectory into adulthood is intrinsically exciting. The variety of choices out there also make it stressful. You could fly across the chasm by going to college, absorbing the abundant wisdom of successful people with online tutorials, or just trust your gut—and spread your wings.

Many active young minds now teach themselves and start their own businesses before graduating from high school. At the same time, many will do as I did and pursue high levels of education to fulfill a deep need to satisfy their intellectualv curiosity and to achieve financial security. Personally, I've enjoyed the benefits of corporate work but also helped empower entrepreneurs in the developing world. Your choices might be different, but I'm here to make sure you have all the tools at your disposal to choose a path that makes you a changemaker.

You don't need to be told that the career you choose isn't just about money. I'm guessing you have already figured out that you want profitable work, a chance to contribute to society, and a satisfying work-life balance. But that trifecta could take innumerable shapes. I wrote this book to make your crucial career decisions a little bit easier while helping you change the world in a way that isn't just a drop in the bucket. It's important because, you see, the world needs you to do this work. You, specifically. You.

It's troubling that we must make life plans even as we become aware that much of life is unpredictable. After all, the extreme weather we've seen in the past decade could affect your safety, out of the blue. Avoiding living in regions prone to droughts, floods, and wildfires will be more difficult for you than for previous generations. So, climate change introduces certain new limitations and emotional stressors to Gen Z, but your technology-savvy generation also enjoys nearly unlimited possibilities in the online world. With the introduction of artificial intelligence, those possibilities have multiplied exponentially. Because your life in social media and video chat provides seamless connections to specialized communities, you're used to building coalitions everywhere you go and organizing around the causes you care about. I think that's great, because to fight climate change and keep this planet livable, you're going to need all the help you can get.

In my young-adulthood in the 1970s and 1980s, my generation suffered from a different type of uncertainty—we faced the Vietnam War and the Cold War while spearheading the fights for women's equality, racial equality, and gay rights. You face upheaval, too, but my experience as an engineer, manager, mentor, and family man has taught me a few tricks about fighting for justice and self determination that I believe transfer across the generational gap. Navigating all that uncertainty and potential upheaval requires a complex and integrated mindset that doesn't just think outside the box. It forgets there ever was a box.

You are in a Race to Save the Earth

Many years after that fateful moon landing, as a young man with a newly minted chemical engineering degree, I visited the Gulf Coast petrochemical corridor and stood on a bridge where I saw mile after mile of intertwining pipes, vessels, and distillation columns weaving over and under one another, extending miles in every direction. At that innocent age, it looked to me like a beautiful 3D tapestry. I understood that this region accounted for half of US chemical manufacturing, petroleum refining, and natural gas processing and was in awe at its power. I even felt that I was truly fulfilling my dream of improving others' lives through science and technology. But within that same year, I became acutely aware of the dire need to adopt environmentally and socially sustainable practices lest our planet's damaged ecosystem cease to support life on Earth. With a growing concern, I read the latest scientific findings, which painted a bleak picture of massive global catastrophe within a century, should we fail to undo the ecological damage caused by generations of industrial growth. To help find a better path forward, I attended the Business & Sustainability Leadership Programme offered through Cambridge University and sponsored by Prince, now King Charles, and even had the opportunity to meet with him twice on the subject of sustainability.

Early in my career, I learned that mining resources from the earth, using them in chemical reactions, and thoughtlessly disposing of the resulting waste was a linear model—a primitive concept, really, providing energy in the short term but destroying Earth over time. That's when I became more interested in a circular model of resource use, where waste of all kinds is recycled, reused, repurposed, or re-engineered in a way that eliminates the notion of waste altogether. So, the next time I visited the Gulf Coast, I stood on the same bridge, gazed out at the same miles and miles of pipes snaking this way and that, all the way to the horizon, and my only thought was, *What a rape of the earth.* I vowed to innovate differently.

Surely, right now, you're smacking your forehead at what a noob I was, in the beginning. Indeed, it took my generation way too long to realize petrochemicals were destroying this planet. But you're already deeply aware of the ravages of climate change, plastic pollution, air and water pollution, deforestation, and the rampant loss of biodiversity. There is, in short, a lot of work to do now, before it's too late. In many ways, human society is in a neck-and-neck race with the factors that negatively impact climate change. At the same time, industry of many kinds keeps our world running, so how can we find a balance? This book addresses that important question and how to work with, not against, current trends in order to build a better world.

Gen Z has seen the 2012 UN Conference on Sustainable Development,[1] the 2015 UN Framework Convention on Climate Change,[2] the 2019 UN Climate Action Summit,[3] and the famous 2015 Paris Accords,[4] where world leaders pledged to prevent global temperatures rising by more than 1.5°C. If that level of international commitment to sustainable development hasn't inspired you to act for the environment, well, perhaps issues around human rights and gender equity are more to your liking. It's reasonably rare that I meet a Gen Zers who cares nothing about these pressing issues, as you've grown up watching scenes of today's rampant wildfires, hurricanes, floods, and droughts on television. Perhaps you've even lived through some of these catastrophic, climate-influenced events yourself.

Then again, US withdrawal from international accords has polarized people in every living generation. So, even though you've seen and experienced climate-related devastation, you may position this in your mind as more of a political challenge than a scientific one.

While recycling and its circular economy were once new to me, they've been part of your life since the beginning. Yet, you probably know that much of modern business and industry is sadly still mired in a linear economy. There is so much more potential you can achieve by re-setting your goals away from that outdated quest for bigger paychecks and higher market share and toward reasonable profits paired with sustainability goals. The way this is done today is so different from the old days, when we left environmental action to nonprofits. Now, motivated young adults like you are taking their own initiative to build startups and join corporations with sustainable global outreach, making doing good part of making a living. This book details both those paths and how technology is here to help you.

In many ways, this book is the culmination of my work attempting to turn mere engineering into true innovation and changing small-scale development into big-picture problem solving. I've achieved a lot, so far, but not enough for my liking. Thus, I offer my guidance as to how you can join me in pursuing the crucial goal of sustainability in every aspect of business. Yes, engineering for sustainability is more difficult and time consuming than the dead-end linear model. It costs money and requires patience. What's more, collaboration and cooperation between industries is essential, which means corporate management and leadership must play a pivotal role in uniting all the elements necessary to create that sought-after circular model of industry.

The Future is Already Here

As I write this, technology is exploding. With the advent of artificial intelligence, 3D printing, virtual reality, genomic testing, and innumerable other novel technologies, the sky is the limit in terms of what the next generation will be able to accomplish. But whether any population gets to reap the benefits of modern technology currently depends upon its nationality, socio-economic status, cultural background, gender, and numerous other inherent factors. Thus, the inconsistent use of technology around the globe is one of today's biggest causes of wealth disparity. Science fiction author William Gibson famously said, "The future is already here. It's just not evenly distributed,[5]" which, I feel, is a pivotal statement for the times we're living in.

Developed nations have access to incredibly powerful technology that will improve and extend our lives, while there still exist on planet Earth groups of people without access to the most basic consumer products. I'm talking about products so necessary that their presence ensures human rights, and their absence mires millions in endless cycles of poverty and isolation. That lack exacerbates exactly the type of economic disparity that leads to war. Let me give you an example.

I've spent the bulk of my career designing superabsorbent and personal care products—everything from toilet paper to diapers to women's sanitary products—as such, I've worked for companies conducting active outreach to developing nations to discover their consumer needs. As a result, my work has become intertwined with world sanitation issues, especially as they pertain to women. I've become intimately acquainted with a disturbing fact: the lives of women who do not have access to modern baby diapers and/or women's sanitary products are significantly restricted in terms of women's access to education, gainful employment, and freedom from abuse. The better women's and babies' sanitary products get (and modern advancements are just short of miraculous) the more freedom women in all societies have to participate comfortably in all aspects of a safe and productive life. Such freedom elevates not only a woman's own status but the entire economy of her region, cultural group, or country. Where those products are not widely available, society falls further and further behind the rest of the world—exacerbating economic disparity.

I'm deeply concerned about ensuring sanitary products are available worldwide, but I'm aware that, despite such availability, many would-be consumers still couldn't afford them. This has led me to ponder the circular resource model. It seems to me: if Earth's resources can be made to function in a way that emulates a natural ecosystem, why can't we develop an economic model that does the same? Such a model would ensure that the user of a product isn't the one paying for it, and yet all parties involved in the product's circular life cycle see some form of profit. It's a complex idea, to be sure, but not impossible, as technology now allows for even soiled, used items to be profitably recycled. Humanity-centric innovation suggests that we develop up-to-date products and an up-to-date, economically viable, recycling system in step with the times.

Pairing Sustainable Innovation with Business

I'm a strong believer in the power of business, more so than philanthropy, to solve world problems like this one in a scalable manner. To do so, in the particular industries where I've worked, we must confront the taboos around talking about sanitation products at all. Doing so for as long as I have has made me aware that there is profit to be made, and important work to be done, in the world of "unmentionables." As such, I want to help you get in on the ground floor of a misunderstood and ignored industry that's ripe for innovation. The notion that recyclable sanitary products could be provided free to the consumer and paid for by the recycling entity is just one example of many you'll find detailed in this book. In fact, I'll discuss many humanity-centric ideas in enough detail that, should you be inspired to take these ideas and run with them, I welcome you to do so. I hope this book scatters the seeds of innovation far and wide.

The engine behind this book and my life in humanity-centric innovation has been a promise I made to myself back in 1997, when I realized that, after years of trying, my wife and I were going to have our first child, who would be part of your changemaking generation. The realization that my life's work could help create the physical world my daughter would inhabit and that my choice of work would likely inspire her own left me with a great sense of responsibility that caused me to sink into days of deep contemplation. I asked myself what I'd accomplished in life, so far, and what I wanted my legacy to be. I've always believed that finding one's true purpose and pursuing it is the key to happiness, so I felt compelled to write down a vision statement to guide my life as a parent. I wrote that my purpose was to "help raise children that live a life fulfilled," and to "help create businesses that improve people's lives." So, while the first *Generation Z* kids were born in the mid-1990s, I was, in a way, reborn that year, myself.

The inclusion of the word "help" in my vision statement felt pivotal to me, as I'm aware that I, alone, cannot achieve any of these things. Both in my business and personal lives, I work with others, collaborate, cooperate, lead, follow, and attempt to inspire. As a husband, father, innovator, and corporate leader, I'm a cog in a larger machine. Professionally, I reserve the right to choose that machine and ensure it's programmed toward goals that resonate with my own. As such, I've made many changes in my work life, over the decades, to ensure I'm always striving for humanity-centric innovation and nothing less. This book is the latest in my efforts to pursue that long-established personal and career goal. I aim to help others see that with technology developing as fast as it is, we've got to consciously bend it towards humanity-centric ideals. Or else, it'll do exactly what many science fiction films have depicted: function in an inhuman manner and destroy the planet to the point where it becomes inhospitable to most forms of life. The choice is ours. At first, the methods for steering technology in the "right" direction may seem mysterious to the uninitiated, but trust me, it's possible. I've spent my professional life doing it. This book is a primer for how you, too, can live a life of innovation—a life fulfilled.

This book is an invitation to journey with me as we explore the transformative power of living a purpose-driven life. It is a testament to the profound impact purpose can have on guiding both life and career choices and inspiring businesses to transcend traditional models of success. I hope that the following chapters inform your decisions, shape your aspirations, and influence the legacy you wish to create. After all, a purposeful existence is the most potent motivator in the world, capable of catalyzing change that shapes the future of the planet.

1

LIVING WITH THE RAVAGES OF
FOUR INDUSTRIAL REVOLUTIONS

The greatest threat to our planet is the

belief that someone else will save it.

— Robert Swan, the first person to walk to both poles and the force behind the
South Pole Energy Challenge—the first expedition of its kind—a 600-mile journey
to the South Pole with his son, surviving solely using renewable energy

First, the good news: we're living in an unprecedented time
in human history, where technology is advancing so fast that
it is, or will soon be, capable of solving all the world's major
problems. Now, some of you may be thinking, *Nonsense! Technology and industry are the cause of most of the world's problems!* And
you'd be right about that. But that doesn't make my statement
wrong. They say ignorance of history makes one doomed to

repeat it, and never more so than in the case of technology. If we're not careful, the danger of repeating the damage caused by early tech is very real, so let's briefly revisit technology's history. You see, 200,000 years ago, the humans that walked on Earth were anatomically similar to you and me, but for the first 130,000 years, we lived like animals. Then, about 70,000 years ago, something happened. No one knows what caused the change, but it's now known as the cognitive revolution or great leap. Archeologists tell us there was an explosion of art, tools, and human migration. Anthropologists tell us humans discovered speech and invented language.[6] Ever since then, each time we've discovered a new way to connect, communicate, and collaborate, our world has fundamentally changed. Just look at the differences in human society brought on by learning to talk, then write, then do mathematics, followed by the social upheaval brought on by inventions such as the printing press, radio, TV, internet, World Wide Web, social media, now generative AI, and soon, agentic AI. Our advancements have put us humans on an exponential trajectory of growth.

The effect of humans on the planet really heated up when, in the late 1700s, we enjoyed the first industrial revolution featuring the invention of the steam engine, weaving loom, and cotton gin. At that time, humans first experienced the miracle of rapid transportation and mass-production and thought the world was getting better in every way. However, as we began to burn fossil fuels in earnest, pollution came along with the progress. This was, in many ways, the beginning of humanity learning to live with that disturbing dichotomy.

It would take approximately another hundred years before the planet's next industrial revolution in the late 1800s, where electricity came into common usage and assembly lines further automated production. People in the Victorian age were in awe of the fact that their homes were lit up well past sundown and thought technology really had reached its peak! But society now needed large, central power stations, pollution increased as industry and automobiles flourished, and dangerous electrical wires snaked across cities and towns, meanwhile people stopped living by the natural rhythms of the sun, which changed society forever. Humans lived with the ups and downs of this technology for about another hundred years.

Finally, in the late 20th Century, around 1970, we built microprocessors, which led to the common use of electronics and enhanced our ability to automate life and business. Boy, did we think we had it all figured out, then! Telephones were small and cheap enough for individuals to have in their homes, and by the early eighties, computers were, too. Cars became safer, more sophisticated, and affordable, further enhancing personal freedom. During this time, we also developed innumerable chemical compounds for mass use in agriculture and industry, maximizing production at every level. At first, we had no idea the irreparable harm some of these substances caused to the planet, and when we finally gained awareness of the devastation our technological progress had caused, there wasn't much anyone could do about it. At this point in human evolution, going backward to a pre-industrial era was out of the question. But with this third industrial revolution, technology itself had reached a stage where it enabled its own rapid development, so it didn't take another 100 years for us to get to Industrial Revolution 4.0. In fact, it took half that time.[7]

As I write this around fifty years after the advent of the microprocessor, we're well into the planet's fourth industrial revolution, where the power of a laptop computer has been harnessed to a motherboard the size of a salt crystal;[8] where cellular phones, "smart" systems, and the internet dominate our technological landscape; and where social media has created societies as big as countries with no physical location, enabling anyone from any social class to share information of any kind with the world at large. The technological developments of the last fifty years have, in fact, turned the very idea of social revolution on its head.

Instead of offering better transportation options, this era enables people to leave home less as they conduct business meetings, see doctors, and even run political campaigns via video instead of in person. Instead of offering a new set of jobs, this era enables self-employment through crowdsourcing such as home-share and ride-share apps. Instead of fighting for better physical working conditions, this era simply enables many to work from home. Meanwhile, much of the work formerly done by humans is now handled by ChatGPT, other generative AI, and various robotic applications,[9] which, in some ways, means this era requires a better-educated population that can handle more managerial-level jobs, but education is easier to get than ever, online. Progress is no longer measured by how we maximize the use of electricity, petroleum, and steel. Indeed, today, data is the new oil; artificial intelligence, the new electricity; and robotics, the new steel. Future companies will either embrace data, AI, and robotics or cease to exist.

Because of the rapid pace of technology, there will be more wealth created in the next ten years than was created in the last one hundred. In fact, according to ARK Investors,[10] the estimated economic impact in the next decade will dwarf the economic activity generated over the last century from well-known industries like telecommunications, automotive, electricity, computers, and the internet. This new wealth will come from blockchain technology, genomic sequencing, robotics, energy storage, and artificial intelligence. This is no surprise, as all wealth, from the beginning of time, has come from innovation – doing something different that creates value. Innovation itself comes from a combination of knowledge, curiosity, imagination, persistence, luck, and the application of the scientific method. Humanity has simply gotten so much better at these skills that technology now even enables us to predict its own progress. We know the next technological revolution won't take another hundred years. In fact, it will probably begin fewer than fifty years from now.

Those of us alive today may very well see the world completely transform, yet again, within our lifetimes. To this point, the former vice chairman of General Motors has gone on record saying that by 2035 nobody will own or operate a vehicle.[11] We'll simply hail self-driving taxis to take us everywhere. Another futuristic company called Neuralink has already developed brain-computer interface technology where electrodes are implanted into the brains of quad-riplegics to enable them to operate computers with only their thoughts.[12] At the same time, a start-up called Atomic Machines has developed technology that reorganizes matter at the atomic level, enabling rapid manufacture of any object of any size in record time. What's more, genetic disease will soon be a thing of the past, as the CRISPR-Cas9 genome editing tool will soon allow scientists to alter any organism's DNA to cure diseases such as inherited blindness[13] and sickle cell anemia.[14] Indeed, emerging technology is set to improve human health in so many ways that children born this year will likely live to be more than one hundred, but only if they have a planet to live on and societies to live within. After all, our planet is still suffering from the ravages of four industrial revolutions.

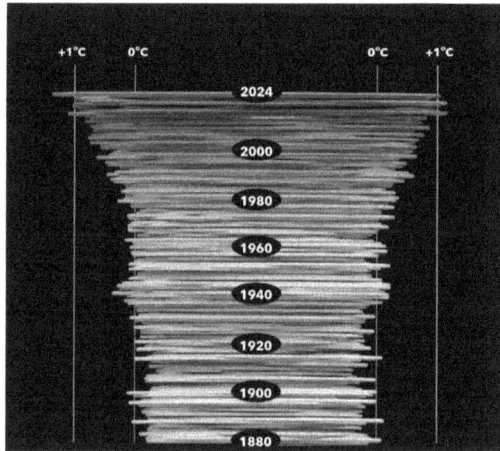

This chart shows data from NASA[15] illustrating how the surface temperature of Earth has changed from 1880 to 2024. And it's still climbing. In fact, Princeton University,[16] The Royal Society,[17] and NASA[18] have all issued reports announcing that even if we immediately stop all fossil fuel use (which ain't happening, let's face it) the CO_2 already in the Earth's atmosphere would continue to warm the planet for hundreds if not thousands of years. So, to stop global warming and preserve the earth before it ceases to be hospitable to human life, we have to figure out ways to *remove* CO_2 from the atmosphere. Just fifty years ago, such an idea would have been laughable, but with the technology we have now, surely there is a way. What's more, with the weaponry and strife between nations currently existing on the planet, there's a better-than-average chance humanity will actually destroy itself before global warming finishes the job. One company, Climeworks uses direct air capture technology to remove CO_2 from the atmosphere, storing it underground through mineralization.[19]

Over time, technology has proven to be both a blessing and a curse, and this is still the case, as it has created a world that now changes so drastically, so often, that life today has been described as VUCA: characterized by high Volatility, Uncertainty, Complexity, and Ambiguity. Keeping up with the way technology constantly updates our tools (and changes society in the process) has practically become a part-time job we each have to do in addition to making a living. It's yet another aspect of technology's blessing/curse dichotomy and the reason technology will either make us or break us, as a species. What's more, because of the speed of change, whatever happens will happen *in your lifetime.*

Defining Humanity-centric Innovation

As a parent, I think it's safe to say that I'm not alone in dreaming that my children's children's children will tell stories about great-grandpa's life and the times in which he lived. You could say my greatest ambition is to be a good ancestor. But will I be? Will our progeny look back and admire my generation (or yours?) for seizing the opportunity technology presented to save a dying planet? Or will tired parents be barely able to choke out their stories through a smoggy haze to children wilting in the latest heat wave, hiding in a bunker from a violent post-apocalyptic society, trying to achieve sustenance on polluted water and unclean food? In fact, asking, "Will we be considered good ancestors," as posed by Roman Krznaric,[20] is a great way to underscore our obligation, indeed our mandate, to save this planet for the generations to come and even to ensure those future generations exist at all. I often wonder: will history textbooks one day describe the eight billion people alive today as those who ushered in a new Renaissance? Or the next Dark Age?

When people act to clean up our oceans, grow chemical-free food, reduce CO_2 emissions, and curb the flow of pollution, they often speak of "saving the planet," which is wonderful, but I think it's important to note this language is the wrong way around. After all, for the most part, planet Earth is only suffering because we human beings are on it.

We're the ones who created the Great Pacific Garbage Patch, which contains two trillion pieces of plastic and is twice the size of Texas. We're the ones who have been dumping planet-altering CO_2 into the atmosphere for centuries, fueling global warming and raising sea levels so rapidly that some island nations are already losing land to the ocean—and, in the not-so-distant future, may vanish entirely. We're the ones who developed the nuclear weaponry that made us aware of a looming phenomenon called "mutually assured destruction." If we humans die out because our societies and atmosphere are no longer conducive to human life, trust me, the rest of the planet will recover.

The earth has been here for four and a half billion years and will likely remain a lush, green oasis for billions more, although humans may not be around to enjoy it. After all, without humans here to cause trouble, the forests, in a couple thousand years will grow trees right up through our buildings' concrete foundations and twine vines around our defunct power stations. Animals will evolve, thrive, and continue to live by instinct, even if radiation gives them three heads each. Fish will eventually evolve to function even better with plastic in their systems, and they'll repopulate the oceans. For proof, look at Chernobyl, where a defunct nuclear power plant makes the place famously toxic to humans but hasn't stopped a woodland from growing over the town and deer from grazing next to the faulty containment vessels.[21] Rest assured, *the planet* only needs saving insofar as it is the container for human life. In fact, if Mother Earth could talk, she would probably say, "The sooner I'm rid of you pesky humans, the better!" Humanity itself is the thing we're striving to save, but first rescuing our planet is the path to that goal. We need to save the planet to save our species!

Become a New Kind of Billionaire

Isn't it incredible to realize that, unlike all previous industrial revolutions, this fourth industrial revolution (which, as I write this book, is just getting started) finally provides us with the tools to improve all the harm we've done to this planet throughout the centuries-long process of developing the tools themselves? For instance, now we have blockchain technology that ensures secure, transparent, hack-free information-sharing across worldwide networks. We can do genomic sequencing, which is leading to genomic medicine, where certain disease markers could be permanently edited out of individuals' DNA. We have even fulfilled, to some degree, the fantasy presented by The Jetsons cartoon so long ago: we have advanced robots that can do everything from vacuum our floors to calculate algorithms that run entire digital societies. What's more, we have access to a new system called additive manufacturing that can (or will soon be able to) 3D-print everything from tiny objects measured in nanometers to a full-sized house in record time.[22] We have long-range batteries that not only enable us to drive electric cars but potentially will provide portable energy solutions for every city, village, and rural outpost in the world. And, of course, we have artificial intelligence (AI), which reduces human error, processes data in record time, and does the dirty work in regions too risky for humans. What's more, part of AI's job is to constantly improve itself, so the more we use it, the better it gets.

But these technological tools are so powerful that, if we don't use them right, they also have the potential to exacerbate our existing problems. The greenhouse effect is only one of the types of environmental damage that needs to be reversed. On top of that, the cells of almost every living organism on Earth now contain micro- and nano-plastics. Studies show fish exposed to such microplastics experience a variety of toxic effects including structural damage to the intestine, liver, gills, and brain as well as reduced fertility in subsequent generations.[23] Unfortunately, micro-plastic toxicology studies on humans are still in their infancy, so we still don't know what these plastics are doing to us. The only thing we can be sure of right now is empirical evidence points to nothing but negative effects. Yet, today's technology gives us the power to find a way to stop and reverse these and many more indicators of man-made planetary failure. But will we?

As Carl Sagan observed, "Extinction is the rule. Survival is the exception."[24] In our current era, it's imperative that we harness modern innovations to restore and preserve our planet. Historically, our pursuit of profit has led to environmental challenges; however, that same drive for advancement can now be redirected towards sustainable solutions. By channeling our collective ambition and creativity, we have the opportunity to repair past damages and build a more resilient world for future generations. Many on this planet still aspire to use and abuse every resource imaginable in a rush to become millionaires or billionaires. Hear me out as I encourage you to do the same, but with a twist. What if we stop defining a "billionaire" as someone who accumulates a billion dollars and instead define it as someone who helps a billion people? Now, that's a goal worth pursuing! Most folks, in all different walks of life, believe their work may help tens or hundreds or even thousands of people in their lifetimes, but consider that if you work for real change, systemic change, in some crucial aspect of the world, you'll be helping future generations, too. If you help fundamentally alter our current unsustainable systems into long-term sustainable ones, then, over the years, long past your own lifetime, the tens, hundreds, or thousands of people you initially helped will grow exponentially. You'll help a billion people. Yes, you.

The key to becoming this new kind of billionaire is in choosing or altering your career today so that it's oriented toward making deep-level change. If you're currently manufacturing widgets that have little use in the world besides generating profit, wouldn't it be more satisfying to manufacture something that solves a big-picture problem for a long-term effect? If you make your living by providing a service that simply puts a band-aid on problems, wouldn't it be more satisfying to work for long-lasting change in the same industry? Even if it's just a change in how people think or how they conduct one small aspect of their lives, if you make it your mission to create fundamental, positive change in whatever aspect of the world you control, you could become a "billionaire" with the legacy you leave behind. I call this "planting a tree whose shade you'll never enjoy."

It has always been my habit to think long-term in that way. In fact, I find it the key to job satisfaction in every way. The days of short-sighted "working for the weekend" really need to come to an end. This suggestion begs the question, *What are the world's most crucial problems, and how do I choose one to focus on, so I can help a billion people?*

In the late eighties, when I first started my chemical-engineering career, I asked myself the same question. Then, I read about the Brundtland Commission, which had been conducting studies on this topic for three years. Its 1987 Brundtland Report entitled "Our Common Future," quoted the Great Law of the Iroquois Confederacy:

"In our every deliberation, we must consider the impact of our decisions on the next seven generations."

This report highlighted the fact that members of high-income countries like the United States use such a disproportionate number of resources that if everyone in the world were to live like us, we'd need an additional 2.6 Earths to support us all.[25] The impact of the report was that we can no longer talk of economic and environmental policy in separate compartments. It introduced "sustainable development" as a bridge concept connecting economics, ecology, and ethics. Its leaders stated, "To get real action, the ownership of the concept of sustainable development must extend to all sectoral agencies and, most importantly, to key private-sector stake holders."[26]

In essence: Don't wait for the government to force you to act sustainably; take it upon yourself. Many nations took this report very seriously, and one of the eventual results was that the United Nations established a Commission on Sustainable Development that defined seventeen Sustainable Development Goals (SDGs) for the planet. This list provides an excellent guideline for anyone seeking direction as to how to make a difference for humanity and the planet. In brief, it follows:

17 United Nations Sustainable Development Goals[27]

1. End poverty
2. End hunger
3. Provide good health and well-being for all
4. Give access to quality education
5. Achieve gender equity
6. Ensure clean drinking water and hygienic sanitation
7. Construct systems providing affordable, reliable, sustainable and modern energy for all
8. Enable all populations to have decent work leading to economic growth
9. Build resilient infrastructure while fostering innovation
10. Reduce inequality in socioeconomic status
11. Make all cities and settlements safe and sustainable
12. Consume and produce responsibly
13. Take urgent action to combat climate change
14. Conserve the sea and use it sustainably

15. Protect terrestrial ecosystems and sustainably manage forests
16. Promote peaceful and inclusive societies with justice for all
17. Revitalize a global partnership for sustainable development

I hope you can find an SDG above that inspires you or dovetails with your existing skills. Shortly after I read the Brundtland Report,[28] number six became close to my heart quite quickly, inspiring me to build my career, as much as possible, around achieving various aspects of this goal and guiding the corporations I worked for to participate as well. If you truly contemplate this list and ask yourself which aspect of world sustainability feels both significant and possible for you to work towards, you've already begun the process of becoming a modern-day billionaire. With this new "billionaire mindset," your efforts toward innovating a sustainable future will progress quickly from hype to hope to something really happening.

The Be-Your-Own-Boss Mindset

In the traditional 20th century model of employment, institutions hired individuals to achieve the purpose of that institution. Employees worked for bosses and performed to the best of their ability, no matter whether they were Rosie the Riveter or Albert Einstein. This still takes place today, and workers' job satisfaction tends to depend upon the quality of the management that directs them to produce whatever they produce for their company. I have worked for many such companies, managed others, and been managed by those higher placed than myself, so while my work model has been, on the surface, traditional, hierarchical, and subordinate to corporate interests, I never thought of it that way, which made a huge difference in my job satisfaction. In my heart, I've always seen myself as my own boss, with the company as a partner in achieving our shared vision, rather than the other way around. I entered the workforce knowing I wanted to make a difference in society, change peoples' lives for the better, and fulfill my childhood dream of using science and technology to build a better world. From my very first job, I've seen each company I work for as a valuable part of my journey, helping me grow and move closer to my ultimate goal. In the companies I worked for, whenever I was given a choice as to which projects I wanted to work on, I'd choose something around sustainability, like engineering systems for clean water or developing projects for world sanitation or women's equality. When I felt one company wasn't helping me fulfill my own goals, I sought work at a different company whose vision was closer to my own. I always asked myself, "how do I get the company working for me?" as a means to achieve my humanity-centric ambition. I think this is more of a 21st century way of thinking. After all, nowadays we must take personal responsibility to ensure that the work we do advances the cause of saving humanity from the possibility of extinction.

You don't have to be the CEO of a corporation to make

that resolution, you just have to be the CEO of your life, create a vision statement that drives you forward, and (as much as possible) don't compromise in the achievement of that vision through your employment, where you use the resources your company allocates toward those goals. I've personally observed that many members of Gen Z are highly ethically driven and can't be motivated by money alone. I think as we move further into the 21ˢᵗ century, we'll see more and more individuals "hiring" institutions to achieve their own ethical goals. Companies that have no sustainability initiatives or outreach to underserved communities are going to find themselves understaffed as qualified workers take employment elsewhere, or create their own companies.

During the current era in my own career, which I suppose would traditionally be called retirement, I've chosen to start my own business that promotes sustainable initiatives all over the world. I call it my renaissance, not my retirement. After years in the corporate sector, I'm finally living my dream to the fullest. But many brilliant younger engineers these days don't see any reason to wait to become self employed when they have so much skill, access to nearly infinite technology, and are living in the midst of what many call an AI revolution. Nowadays, in many cases, it literally makes more sense to start one's own business than to get a job working for others. What better way to begin helping a billion people? In fact, dozens of such start-ups are so successful, they have been valued at $1 billion before going public.

Sam Altman, CEO of OpenAI, has given startups like these his stamp of approval, further speculating that with AI taking over repetitive-task and data-analysis jobs, he expects to soon see the first "one-person unicorn," or billion-dollar company operated solely by its founder.[29]In my youth, the world's greatest technological tools were only available to major corporations that had massive financial resources, but today, nearly anyone with the skill to use them (or the desire to learn) can get any kind of advanced tech on a laptop. This fact alone disrupts traditional business models and will soon democratize the entire entrepreneurial landscape. I think it's even safe to say the days of having to pay your dues to corporate America are over.

Also a vestige of the past are typical e-commerce business models and other such startups without a social conscience. Just in tech for the money? These days, you'll find a lot less investment capital coming your way. Renewable energy, carbon capture, plastic recycling, HVAC efficiency, and fighting the new rash of wildfires around the world are the types of sustainable innovation topics winning prizes, gaining investors, making the news, and garnering profit, too. Both Simon Stiell, executive secretary of the UN Framework Convention on Climate Change, and UN Secretary-General António Manuel de Oliveira Guterres[30] agree. They have asserted that humanity is on a "highway to climate hell," and since we still don't have the technology to cool the atmosphere quickly enough to save the human race, we must put ourselves in the hands of today's youth via paradigm-shifting startups.

Such startups operate by leveraging tech, innovation, and minds brimming with creativity, drive, optimism, and admittedly a touch of sheer desperation. For such revolutionaries, data is a new source of value that must be mined with the enthusiasm of a startup founder in Silicon Valley. Accordingly, all boundaries must be opened, and no source of this precious resource can remain taboo. Importantly, everyone from UN officials to tech moguls have stopped calling on charities and non-profits to do this kind of data-based, humanity-centric innovation. You see, where sustainability-minded folks used to want to "stick it to the man" and work outside the corporate sphere, things have changed. Now, making a profit is no longer synonymous with exploiting people and resources. Quite the opposite, in fact.

In a way, the global polycrisis has done the world a favor by excising that victimhood tendency a lot of young folks (well, most of us at some point in our lives) used to carry around. Once upon a time, it was, "If only my boss would tell me what to do, I'd do it!" but now the prevailing attitude among educated and capable folks like you is, "Someone must solve this problem. Why not me?" This is an especially powerful statement when it becomes clear you can now make a living, often a very good living, pursuing humanity-centric innovation. To be completely fair, over time, Non-Governmental Organizations (NGOs) and nonprofits have played crucial roles in making an impact on nearly all the UN SDGs, but without significant profit, they just can't scale up enough to help more than a small sector of any population at a time. But with the planet now in the midst of its fourth industrial revolution, we don't have time to play small ball anymore. Go big or go home.

2

THE FUTURE OF
SUSTAINABILITY IS BUSINESS

Whether you believe you can do a thing or not, you are right.

— Henry Ford, an industrialist and pioneer in making automobiles
affordable for middle-class Americans

As a lifelong technologist and businessperson who has always had sustainability top-of-mind, I believe in the power of profitable business to make a difference, but I also know it must be part (or most) of the business' stated vision, not just a recreational side project. There is no shame in building a profitable brand around providing clean water to underserved populations, manufacturing earthquake-proof housing, empowering women and girls in developing nations, or cleaning plastic out of the ocean.

This idea is best illustrated by the following graphic, which I call the "sustainability maturity curve:"

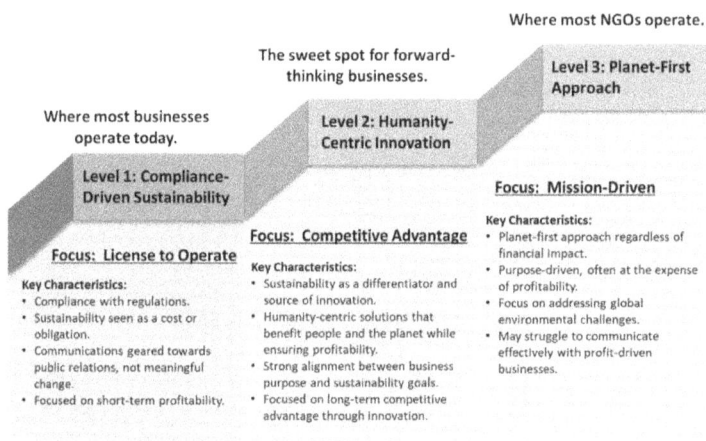

Where most NGOs operate.

The sweet spot for forward-thinking businesses.

Level 3: Planet-First Approach

Where most businesses operate today.

Level 2: Humanity-Centric Innovation

Level 1: Compliance-Driven Sustainability

Focus: Mission-Driven

Key Characteristics:
- Planet-first approach regardless of financial impact.
- Purpose-driven, often at the expense of profitability.
- Focus on addressing global environmental challenges.
- May struggle to communicate effectively with profit-driven businesses.

Focus: Competitive Advantage

Focus: License to Operate

Key Characteristics:
- Sustainability as a differentiator and source of innovation.
- Humanity-centric solutions that benefit people and the planet while ensuring profitability.
- Strong alignment between business purpose and sustainability goals.
- Focused on long-term competitive advantage through innovation.

Key Characteristics:
- Compliance with regulations.
- Sustainability seen as a cost or obligation.
- Communications geared towards public relations, not meaningful change.
- Focused on short-term profitability.

At the lowest level of the curve, companies operate by employing the minimum sustainability initiatives required by law. At the highest level, NGOs operate purely to do what's right for the earth and future generations with no thought to profit or the economic feasibility of their organization. These two extremes often fail to effectively communicate with one another. In the middle, however, is a third level where a for-profit business pursues sustainability because doing so actually creates a competitive advantage in the marketplace. This is where I place humanity-centric innovation. Even so, as I'll explain throughout this book, today's sustainable entrepreneurs often must develop alternative business models that pull profit from unexpected places, never from the true victims of climate change, natural disasters, or social unrest. Thus, humanity-centric innovation requires an innovative approach to doing business.

In case you have any doubt that solving climate change in time is indeed possible, let me explain why now is not only the *last* possible moment to innovate for humanity, but also the *best* possible moment. In the run-up to the industrial revolution we're currently experiencing, the world changed in some profound ways that might actually create opportunities for positive change that didn't exist in the past, but only if we seize the opportunities presented.

First of all, the world is digitizing at a mass scale. In fact, the top ten populations in the world today aren't countries but social media platforms. If Facebook were a country, its population would be greater than China, and its GDP, greater than that of the United States. Statistics illustrate that in developed nations, more than half of most citizens' lives are spent staring at a screen, whether this be their smartphone, computer at work, or TV at home.[31] Seventy-seven percent of meetings are now virtual,[32] and, last year, more than half of US adults saw doctors via telemedicine.[33] Additionally, digital devices track our steps, heartbeats, and even emotions.[34] It's a change that has made many declare the world is becoming "smaller" because we can reach out to people all over the globe, and even gather their data, with very little effort. But some of the world's big recent changes aren't quite so obvious.

The makeup of Earth's population is changing profoundly, too. These days, fewer babies are being born in both developed and developing nations. The reasons include technological, environmental, and cultural reasons. But no matter the cause, statistics tell us that we have reached "peak child,"[35] and soon, the global population will plateau, probably at around ten billion people. Then, it'll start to slowly decrease. At the same time, life expectancy all over the world is going up (even accounting for things like child mortality, war, and disease). In 10,000 BC, human life expectancy was thirty-one years. Fast-forwarding to 1900, it was still only thirty-four—not much of a change. But over the last one hundred and twenty years, life expectancy more than doubled to seventy-one. At this rate, we can realistically expect today's toddlers to live to be centenarians. In fact, globally, centenarians are one of the fastest-growing demographic groups.[36] It all adds up to one puzzling statistic: there are currently more people over the age of sixty-five than under the age of five. There are more grandparents than grandbabies and this will be true for at least the next one hundred years. How will this affect global warming, pollution, and world peace? We don't yet know, but it stands to reason that creating technology and policy with these facts in mind would be wiser than not.

Interestingly, while the world's population isn't growing, it is urbanizing, and that means a building boom. The United Nations estimates that between 2024 and 2060, we'll double the number of buildings on the planet. That's like building one New York City per month for the next forty years.[37] If you want proof, consider the fact that between 2011 and 2013, China used more concrete than the US used in the entire 20th Century.[38] This fact, on its own, is neither good nor bad. After all, irresponsible urbanization can produce massive pollution, but urbanization is also a great way to create some of the world's most futuristic sustainable communities. It's up to us to make changes like these work for, not against, the planet.

Human and animal bodies are changing, too. As I've mentioned, our food, beverages, and air now contain nano- and micro-plastics to the point where they're saying plastics will be the asbestos of the 21st Century.[39] A new condition, known as plasticosis,[40] is even being observed in animals, where ingested plastics cause scarring in digestive tracts—a sign of just how deeply plastics are infiltrating living organisms. Meanwhile, the highly successful mass-production of food has resulted in fuller bellies, globally, but lower-quality nutrition. In fact, in 2023, more people on Earth died from diabetes than wars, crime, and suicides put together. While starvation remains a tragic reality in many areas, far more people today are dying from the effects of overeating combined with malnutrition—a deadly mix of excess calories and missing nutrients that's fueling a global health crisis. Along with these social and physical changes in our bodies, the explosion of technology that seems to have no end in sight is transforming our world into something that, twenty years from now, we'll barely recognize. Moore's Law[41] (named after Gordon Moore, co-founder of Intel, in 1965) explains this phenomenon to some degree. It states that the cost of computing will decrease as its power increases. Technology's growth over the past five decades shows this to be true, as it has been exponential, meaning as soon as computing power doubles, that doubled amount doubles again in the same time frame, and so on. This remains true, today. A computing system, operated via code, that used to fill an entire room, has, over the course of fifty years, become a machine that average householders can afford, carry in their pockets, and operate with the mere sounds of their voices.

In January of 2017, Google[42] and then Intel announced they had achieved "quantum supremacy," meaning their quantum computing hardware could quickly solve problems no classical computer ever could. This was the height of all computing excellence until ten months later, when IBM announced its own superior program that completely raised the bar for quantum supremacy. A few years later, the University of Science and Technology in China announced it had a quantum computer ten billion times faster than those of Google, Intel, or IBM.[43] On December 10, 2024, Google's Willow achieved in just five minutes what today's fastest supercomputers would require approximately 10 septillion years to complete—equivalent to 10 followed by 24 zeros: 10,000,000,000,000,000,000,000,000 years.[44] And the race to the absolute apex of quantum computing continues. In short, rest assured that computers are becoming exponentially faster and smaller at an unprecedented rate. This type of rapid technological change is likely to create more wealth in the next ten years than all the wealth created in the last one hundred.

The Human Genome Project[45] is a great example of recent exponential tech progress. Beginning in 1990, it cost three billion dollars and took thirteen years to partially map the genome of one human being. Now, members of the general public can get the same information, customized to oneself, in three weeks, for $125. With science advancing so quickly, and keeping in mind Moore's Law, it's entirely possible a whole raft of new life-saving therapies will be widely available and affordable by the next industrial revolution (which may indeed be a medical-science revolution) fifty or fewer years from now.

Blockchain, cloud computing, and artificial intelligence (AI) are three more technologies that have only recently become available to consumers but, hand-in-hand, are now contributing to each other's exponential growth. Together, they empower the world to analyze vast amounts of data at record speed while simultaneously fortifying security. This means users can extract trends and patterns from massive collections of otherwise random data, which leads to more informed and efficient decision making in record time. Vast AI- and blockchain-enabled computing networks offer increasingly personalized experiences because of their ability to analyze user preferences and tailor content based upon existing data. But I have noticed that much of the world's harvestable data is still not being collected. This is often due to social taboos around biomaterial and bodily waste. Since all personal data means potential profits for its producers, companies that reach out to shatter those taboos are the ones with the most potential to profit while economically empowering people of all social classes.

The "internet of things" is a new business technology you've probably never even tried to live without, but back when you were just a toddler, it quickly changed everything about the world. Saving space in your actual home, the internet puts a lot of your "things" in the cloud. Documents that used to exist on paper, music albums that used to exist on vinyl, books that used to live on shelves, video that used to require a television, and research that used to require a brick-and-mortar library. It's all in the cloud, now, working toward humanity-centric goals by preventing waste and resource use. Innovative, sustainable business can thrive in the cloud because, first of all, it's a lot easier and more cost-effective to reach potential consumers via the internet, where one can take payment for goods, as well. Secondly, selling digital products instead of actual ones is a viable, low-cost business plan that provides many innovative products with the potential to positively impact peoples' lives. Over the last twenty years we have realized that the ultimate product is one that does not physically exist, although its function does. Just open your smartphone and look at all the apps that used to be physical objects, such as your compass, map, clock, newspaper, camera, etc.

Corporation vs. Small Business vs. Nonprofit: Why Scale Matters

Historically, big business has viewed green government regulations as burdens detrimental to their bottom lines. But in 2009, the *Harvard Business Review*[46] concluded a study of the sustainability initiatives of thirty large corporations over time. Surprisingly, it revealed that when the largest corporations wholeheartedly embrace green regulations, they actually profit handsomely due to their ability to scale such projects at a multi-national level.

When companies view compliance as a necessary evil instead of an opportunity, they tend only to meet the minimum green requirements, and the idea that sustainability efforts reduce market share becomes a self-fulfilling prophecy. But when forward-thinking multinationals tackle green protocols with the same zeal as profit-motivated innovations, they tend to apply the most stringent protocols, and the HBR study shows these efforts result in significant gains in market share as well as a reputation for leading the cutting edge of an industry. It's all about scale.

For smaller companies and nonprofit institutions, conforming to new regulations may require a nearly impossible capital outlay. Multinationals, however, not only think big but they think long-term. Adjusting to new green protocols can be costly in the short term, but research shows that strategic investments in sustainability can yield significant long-term financial returns, with some initiatives providing returns exceeding 10 times the initial investment.[47] While small businesses often need quick returns, larger corporations can afford to invest in long-term sustainability strategies. However, both small and large businesses play critical roles in advancing sustainable practices, with smaller firms often leading in innovation and agility. When, in 2002, U.S. automakers resisted new sustainability mandates, meeting fuel and emission standards at the minimum required levels. Foreign automakers, who fully embraced the concept, easily outpaced them. By contrast, when Hewlett-Packard was forced to stop using lead solder due to the product's toxicity, the company took the opposite approach and used the opportunity to innovate superior solders that not only complied with existing US laws but also anticipated upcoming European Union (EU) hazardous substances restrictions, so that when the new EU protocols were put in place, HP was already effortlessly leading its industry, worldwide.[48] Now, that's the kind of innovative attitude the planet needs.

When it comes to businesses "going green," advocacy groups and society at large often take a skeptical view against big multi-nationals. After all, there's a widespread belief that while smaller companies operate under the radar, big conglomerates are the real waste machines, driving pollution on a massive scale. Historically, this has been true in many cases, but there is another side to that coin: smaller companies are also less able to invest what's needed to massively transform their operations. In that case, they tend not to overhaul their operations but rather manage separate logistics for each of their markets as minimal changes are put in place, one-by-one, which, in the long run, is expensive and tedious. But when multinationals choose to conform to the ecological gold standard with massive worldwide sustainability initiatives, they benefit from economies of scale and optimized supply-chain operations. Thus, the tables are turned, and "the bad guys" become the green innovators. As the esteemed business author Tom Peters[49] says, "We need to outsmall the bigs and outbig the smalls," striking a balance where we leverage the agility and innovation of a small business while harnessing the resources and impact of a large corporation.

What's more, the tendency for large multinationals to have extensive research and development branches presents them with enormous opportunities to preempt upcoming regulations and even one-up them with green innovations of their own. For instance, Proctor and Gamble's R&D team discovered massive amounts of energy were being unnecessarily consumed by household washing machines. When the company responded by developing effective cold-water detergents, it saved Americans eighty billion kilowatt hours of electricity and reduced carbon dioxide emissions by thirty-four million tons.[50] Meanwhile, the investment in developing new detergents paid off by creating a new, scalable income stream for P&G. Similarly, when Hewlett Packard made the decision to be a green innovator, it launched a massive R&D initiative to investigate worldwide regulatory trends and learned that Europe would soon require electrical and electronic hardware manufacturers to pay for the cost of recycling their products. With both cost-savings and cutting-edge innovation in mind, the company pre-empted the regulations by building a European Recycling Platform that works with more than 1000 companies in thirty countries to scale up that recycling operation.[51] This global effort nearly halved the cost of a similar small-scale recycling program while exponentially decreasing electronic waste. Such is the power of a sustainably (and economically) motivated multi-national corporation.

FedEx is another corporation that profited from the decision to far exceed green regulations. One of its initiatives involved developing a set of software programs that optimize aircraft schedules, flight routes, and the amount of extra fuel on board.[52] The company also built a massive solar energy system and switched to hybrid vehicles. But Fedex's most well-known innovation has been its merger with Kinko's, which enabled the shipping company to offer customers the option of sending documents cross-country electronically instead of physically, then printing and shipping them locally to their destinations.[53] The fuel savings has been massive, and the economic savings is passed on to customers: a sustainability win-win.

For multinationals, smaller companies, and nonprofits alike, changing paradigms requires questioning implicit assumptions behind current practices. It was once thought that people couldn't fly, breathe underwater, or travel vast distances with ease. Questioning those beliefs led to the innovations that made our world what it is, today. This generation's questioning is about scarce resources. Can we grow rice without water? Make a waterless detergent? Develop packaging that grows trees instead of destroying them? Produce energy without a power grid? Profit from waste? Travel without polluting? Achieving such "impossible" goals means going forward from where we, as a society, are right now, using the most advanced technology available and the biggest sets of resources at our disposal. Most of all, it means training our minds to see the big picture in every small-scale problem, the long-term results of every social change, and the downstream effects of every waste product we produce. There is certainly still room for small-time incremental change, but to truly harness the innovative power of business, it helps to think big, start small, and scale fast. In the end, it's all about velocity – speed *and* direction.

That said, as a veteran employee of some of the world's biggest multinational corporations, I've noticed how, in such large companies, teams and departments often operate in silos, limiting cross-functional collaboration. For example, a company may have departments in Korea, Australia, and South America, all working on the same project, but even though they're part of the same company, they might as well be in three separate companies for as much as they interact with each other. Multinational corporations often struggle with fostering radical collaboration, both internally across departments and externally with partners, which is crucial for driving the transition to a sustainable and circular economy. I still believe large corporations have the economic power to drive positive change, and with a more collaborative international infrastructure, they can accelerate growth and make a greater impact. That said, one big advantage still provided only by large corporations is the budget to sponsor globally aware startup incubators that promote sustainability, even in potentially competing companies. (For a list of small-business incubators, see Appendix B.)

As more small, modern startups build their business plans around humanity-centric ideas, they will embrace advanced technology from the start. Unlike large corporations, these emerging companies won't have to undergo an expensive transition to "go green." They simply start out that way. Thus, in many ways, both today's largest and smallest companies are the ones on the cutting edge of humanity-centric innovation.

AI-Based Startups are Poised to Make a Difference

These days, innumerable entrepreneurs are jumping on the AI bandwagon to make their fortunes and do good in the process. When it comes to developing sustainable products, startups have a distinct advantage over established companies because they can integrate modern technology from the ground up. Unlike large corporations, which face costly and complex retrofits to make existing products more sustainable, startups can design with sustainability in mind from the start. Today's innovative sustainable entrepreneurs range anywhere from early-stage AI consultants and educators all the way up to high-impact start-up CEOs building sustainably focused AI automation. SAAS (software as a service) and other subscription-based companies tend to dominate this popular field.

SAAS Startups

Amplify,[54] *Futureproof*,[55] *Persefoni*,[56] and many more similar SAAS companies provide software that automatically analyzes a given company's pollution footprint and provides suggestions for improvement. Such companies also publish sustainability benchmarks crucial to the company's ESG (environmental, social, and governance) reports as well as their CSR (corporate social responsibility) initiatives and KPI

(Key Performance Indicators). For companies committed to doing right by the planet (and whose stakeholders demand it, as well), it's important to get a clear, data-driven picture of how the business is performing, take action to reduce carbon emissions to net zero, and achieve important green certifications. These subscription-based software applications conveniently analyze progress toward company goals, create impact reports and submit sustainability statements.

Having these complex metrics automatically generated and analyzed regularly enables CEOs in all industries to be confident their environmental-impact claims are data-driven truth, not greenwashing. What's more, such AI data-analysis software enables a company's sustainability initiatives to be centralized and monitored on a single platform. This convenience empowers companies to hold their head high as they stay fully on board with humanity-centric goals despite all the typical distractions of a growing business. But SAAS isn't the only AI-based game in town.

AI-Automation Startups

Sustainably focused subscription-based AI businesses run the gamut across industries. For instance, *Haystack* provides an AI-driven soil-analysis service for agricultural growers adopting regenerative agricultural practices with a view to sequestering carbon.[57] After all, it's one thing for farmers to employ sustainable practices such as no-till agriculture, rotational grazing, and cover cropping, but to take advantage of agricultural soil carbon markets, a cost-effective and scalable method for measuring and verifying changes in soil organic carbon must exist to accurately detect such a farm's progress from season to season. In fact, expensive, annual testing can even negate the financial upside of such endeavors for growers. *Haystack* removes this barrier with a highly accurate, affordable, subscription testing service that keeps farmers in the know with AI-generated scientific soil analysis.

Voltpost is another up-to-the-minute subscription startup revolutionizing the use of electric vehicles in cities. The company creates a secondary use for ordinary lampposts by adding EV charging plugs to each one, making them available to subscribing users for reasonable fees.[58] By utilizing already-existing electrical systems, this reduces the city's expense and the construction inconvenience associated with installing standalone chargers. *Voltpost*'s design includes up to four chargers per lamppost and supports additional connectivity, making the design and subscription model future-proof and agile enough to respond to ongoing changes in EV technology.

A truly unique startup called *Sortile* innovates by providing a subscription-based software specifically for textile recycling operations.[59] Ninety-two tons of textile waste is being generated each year,[60] and much of that is because recycling textiles requires sorting them by fiber content, which up until recently has been difficult to determine. The knowledge that only 1 percent of textiles is recycled, and that massive textile dumps exist all over the planet, inspired *Sortile*'s founders to develop software that instantly analyzes the composition of fibers such as cotton, polyester, acrylic, wool, and many blends so that the items can be easily sorted for recycling. Utilizing the multi-functionality of AI, the software also gathers insights on color, traceability, and environmental impact, further empowering textile recyclers to keep these used products out of landfills and give them new life.

But not all startups utilizing AI-based data analytics provide subscription services. Many work directly with corporate and governmental clients in unique situations to specifically predict, ameliorate, or prevent disasters, pollution, and other impacts of climate change on humanity.

Other Innovative AI-based Startups

The German company *apic.ai* helps prevent environmental disasters before they start by using AI, one of humanity's most innovative technologies, to monitor honeybees, one of humanity's oldest and most important species.[61] Honeybees are crucial to human survival. They pollinate one in three bites of food we eat, including 75 percent of flowering plants and 35 percent of the world's crops.[62] But a variety of factors have put their population in decline, and many subspecies are at risk of extinction. While saving honeybees is of paramount importance, this isn't so easy. After all, how are humans to know when their behavior threatens bee populations? German startup apic.ai was founded to solve this very problem.

Apic.ai has developed AI that monitors the behavior of bees in their hives. When a hive's pollinating region is exposed to particular chemicals or toxins, this company's computer vision technology provides comprehensive analysis that ultimately reveals even the slightest effect such changes have on the bee population. Such ecotoxicological studies enable farmers to work towards their own enhanced profits and better long-term land use by selecting agricultural enhancements that won't undermine them in the end by reducing the effectiveness of their most reliable pollinators. This simple idea, which influences important farming decisions, is made possible with a combination of today's most advanced technology and the humanity-centric focus typical of today's enthusiastic startup founders.

Another startup, *EarthScan*, was founded to help people work with and around the reality of our changing environment.[63] With the understanding that climate change is inevitable, this company uses predictive data modeling, machine learning, and state-of-the-art climate science to provide risk-mitigation analysis of climate threats to any business' location or potential location, including acute and chronic hazards.

EarthScan's technology can assess climate hazard probabilities anywhere on the planet using baseline insights that analyze climate trends from 1970 to 2100. This unique technology encompasses the seven primary climate-related risks: heat stress, drought, wildfire, flooding, and extreme wind and precipitation. Seeing into the climate future like this enables companies to select appropriate sites for real estate, industry, and hospitality endeavors. Meanwhile, those companies already invested in locations deemed risky can use the information to prepare for impending disasters years before the local weatherman sends out an all-points bulletin.

This startup takes an interesting viewpoint on climate change with technology that emphasizes the fact that our planet is on its own trajectory, and weather events are going to happen whether we like it or not. To ensure a future for humanity on this big blue marble, we've got to be smart and work around the damage with modern technology's nearly miraculous ability to conduct predictive modeling.

Elipsis Earth is another example of a startup using machine learning and AI data analysis to help governments and local organizations work together for the health of communities and the planet.[64] The company's drone technology scans regions for litter and instantly analyzes the type, brand, weight, and recycling capabilities of each day's typical litter over a specific region, providing municipalities with targeted, cost-effective, solution-specific information.

For instance, a UK coastal resort suffered from excessive litter and was also aware that because of its erosional coastline, much of the plastic waste was being washed out to sea where it would eventually disintegrate into the microplastics polluting oceans worldwide. Aware of its contribution to litter, McDonald's corporation funded an *Elipsis Earth* analysis that would enable it to cooperate with the municipality to make paradigm-shifting changes. Analysis rapidly identified forty-seven categories of litter over 475,000 square meters and documented the performance of the three hundred and fifty trash bins already existing in the region.[65] Studies have shown that McDonald's packaging is a significant contributor to fast-food litter, with a 2009 Keep Britain Tidy survey finding that it accounted for 29 percent of such waste on British streets.[66] The *Elipsis Earth* data was even able to detect the times and locations of maximum litter activity. Data also showed that citizens were doing their best to help by piling litter up next to overflowing bins, but the city's collection was simply falling short.

With this data, the city was able to encourage cleanliness with strategically placed cigarette bins, nighttime glow-in-the-dark bins, brightly colored bins in low visibility areas, and even "disco bins" that targeted evening partygoers. This cooperative initiative reduced littering by 75 percent overall while data even demonstrated a "multiplier effect" showing that an initiative to reduce one particular type of litter encouraged people to reduce all litter to help clean streets remain clean. McDonald's litter was found to be only 1.5 percent of the overall liter, but true to its corporate cleanup initiative, the company responded with a further commitment to reducing high impact packaging. Today's AI and the innovative startup founders who make good use of it enable such data-driven approaches to a cleaner environment to be easier than ever.

Initiatives like *apic.ai*, *EarthScan*, and *Elipsis Earth* show, above all, that people want to do right by the planet, and the only thing often holding them back is a lack of data that would enable them to pinpoint their actions. These examples and many more of today's inspiring startups show that AI technology pairs perfectly with the entrepreneurial attitude of ambitious and conscientious young professionals eager to make the world a better place while also making names for themselves with paradigm-shifting innovations.

3

IT ALL STARTS WITH A VISION

Twenty years from now you will be more disappointed by the things that you didn't do than by the ones you did do. So, throw off the bowlines. Sail away from the safe harbor. Catch the trade winds in your sails. Explore. Dream. Discover.

— H. Jackson Brown Jr., author of *Life's Little Instruction Book*

I paraphrase General Omar Bradley when I say, "It's important to navigate by the stars, not by the light of every passing ship."[67] After all, life is full of shiny objects, and they can really distract us and cause us to bounce around in all different directions. But with a fixed vision in mind—a North Star, if you will—we'll stay on course. In that case, there is no limit to what each of us can accomplish. That North Star is more than a goal. It's a *vision*. For each of us, the North Star we choose to follow sets our destiny.

Vision is the key to any innovative concept. It's not enough to be creative and think differently. You've got to have the courage to envision a completely different future than the one our planet is heading towards right now. With such a vision, you can set wild and lofty goals truly worth your effort. In fact, your tip-top goal might be laughably absurd, but when you break that down into smaller sub-goals, things start to look more possible. And when you break those sub-goals down into the tasks required to achieve them, the whole project comes down to Earth and into focus. After all, that's how lofty goals are achieved—one task at a time.

At one point, it seemed absurd to imagine a railroad system spanning America, connecting far-away regions to enable easy cross-country travel, even over mountains and through conflict-ridden regions. It was a crazy idea, but it eventually worked, because innovators broke the whole system down into individual tracks, then sections of track, and hired workers to lay the tracks, one at a time.

When Roald Amundsen envisioned becoming the first Antarctic explorer to reach the South Pole, he seemed like a madman, but he broke his impossible vision down into sections of the journey, found financial backers, developed a perfect ship for crossing the ice-filled Antarctic waters, and compiled a worthy crew. He tackled each smaller goal by achieving the relevant tasks and eventually became the first to plant a flag on the pole.[68] He didn't get there by "hoping for the best" or "winging it," but by goal setting, task-achieving, and imagining everything that could go wrong (including a polar bear attack!) in order to pre-empt it with a solution. Amundsen understood a visionary is only as good as the complexity of his or her plan.

I once had the opportunity to meet the chief designer for Tesla, who told me a story about Elon Musk. By 2002, Musk had sold two major financial-services companies, Zip2 and PayPal, and had around $200 million burning a hole in his pocket. He wasn't content to just keep doing the same thing over again, nor did he want to retire from business. He was a young man, seeking a vision to drive the rest of his life forward. He invited a group of friends and held a brainstorming session where he asked them all for ideas on where to invest the $200 million.

The conversation went deep into the night, covering a variety of topics and ideas—each more outrageous than the last. Finally, Musk approached their whiteboard and drew a large circle with a small square inside it and announced, "this is what I am going to do!"

When asked what his sketch represented, Musk announced, "I'm going to put a greenhouse on Mars as a lifeboat for humans."

Now, that's a vision. It's suitably inspiring and appropriately impossible. The only way to achieve it is to keep on thinking big. He broke down the project into three subgoals: He'd need to build rockets, of course. But to power such long-range rockets, he'd need the best batteries in the world—energy storage the likes of which the world has never seen. With the sun being the only common touchstone for both Earth and Mars, it seemed logical that those batteries should be solar powered. With these goals in mind, he went on to found SpaceX to develop rockets for private space travel, Tesla to develop advanced batteries, and Solar City to maximize the power of solar energy.

Along the way, Tesla has created rechargeable, lithium-ion batteries that are most-famously used to power Tesla vehicles. SpaceX has launched rockets such as the Falcon 1, which became the first privately funded, liquid-fueled rocket to reach Earth's orbit. Solar City has built some of the most efficient solar panels on the market and joined forces with Tesla to build solar-powered batteries to fuel homes. Musk may or may not achieve his grandest vision within his life-time, but along the way, that vision has spawned innovation.

With that in mind, I ask you, future entrepreneur, poten-tial visionary leader: what's *your* Mars shot? Do you have a vision that's both burning with importance and deliciously impossible? Which aspect of our beautiful but imperfect world would you like to improve? What type of innovation would best use your unique skills and strengths?

After all, if you don't think about your future in a vision-ary way, considering how your existence might affect the big picture for humanity, you'll find yourself much like Alice in Wonderland[69] when she asks the Cheshire cat, "Would you tell me, please, which way I ought to go from here?"

"That depends a good deal on where you want to get to," said the cat.

"I don't much care where—" said Alice.

"Then it doesn't matter which way you go," said the cat.

If we ignore the fate of our planet and humanity's future, then setting a vision becomes meaningless. But when we choose to care about something greater than ourselves, we unlock a deeper sense of purpose—one that can drive real change and create a legacy that matters. My life has proven to me, time and time again, that when people seek their true purpose in life, the entire universe seems to conspire for their success. For young folks, it's easy enough to go out there and get a job that utilizes your innate and acquired skills, make a living, and consider that a life. And for established businessmen and women it's perfectly socially acceptable to keep chugging away at the profession where you've found financial success. Human activity is rapidly pushing the planet toward conditions that threaten our own survival. Only humans can solve this problem, and we have no time to spare in trying. I urge you: *Rise to the challenge and lead the way*. If not you, who? If not now, when?

Technology has handed us incredible tools with which to achieve the necessary miracle, so the only question remaining is: *will we?* If each person on Earth today develops an aspirational, important, and driving vision for saving some aspect of this troubled planet, no matter how large or small, humanity really does have a chance. If not, we're just playing with this big blue marble like it's a broken toy we're simply using up before throwing away. Setting your personal vision is the key to changing all that, the key to navigating by the stars.

I want to reiterate that being ambitious is the key to a vision that works. Don't let anyone tell you to be cautious or practical when setting your vision. Musk certainly didn't!

A good way to ensure you're taking charge of your life with just the right amount of ambition is to think in terms of "the cone of strategic possibilities."

The Cone of Possibilities

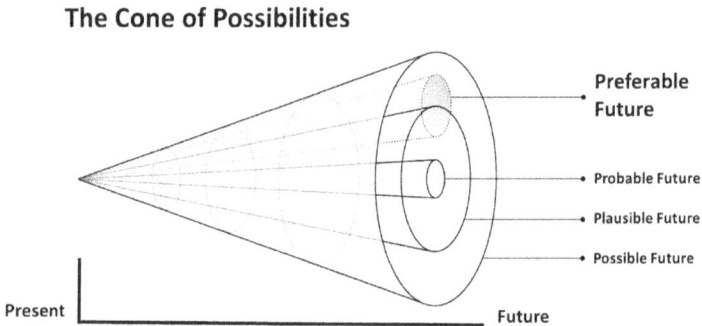

You'll notice that, in the image, when your eye travels from left to right, this represents the passage of time in your life. Whether you're a young person just starting your professional life or an established professional reinventing your life, you'll be starting on the left, where the only thing clear is that you'll eventually wind up somewhere in the cone of the possible future, which encompasses all the others.

If you head in a straight line, never questioning the status quo, you're likely to end up right in the center cone, that of the probable future. It's what's expected, after all. Furthermore, the diagram indicates that what's probable lies within the field of what's plausible (realistic and believable). So far, you're just riding the wave of life, going with the flow.

To live a life inside the shaded ellipse of a preferred future, you're going to have to diverge from the straight and narrow path. You'll have to consciously list the options of what's plausible for you, also look at what's probable considering your particular talents, and then create a preference that unites the possible, probable, and plausible into a consciously chosen preferred future. There is little chance you'll end up in the cone of the preferred by accident or happenstance. Yet, at the same time, it's easy to drive your life in that direction. All you need is a clear vision.

Vision, Value, and Branding

I've talked about how, from an early age, I have felt compelled to make the world a better place through science and technology; yet I eventually realized that motivation wasn't specific enough to be a true vision. I didn't set a vision until I was thirty-six years old. You see, my wife and I were eager to have a family but were struggling with infertility. Then, one Sunday in December 1997, I went to a Catholic church that was not my usual church. There, I heard an inspiring homily based upon a story called "Where Love Is, There God Is Also," by Leo Tolstoy.[70]

In the homily, an old cobbler expected Jesus to visit him, so he prepared a huge buffet, but the dinner hour came and went, and nobody arrived. Around dusk, an elderly man arrived to beg for food, so he fed him from the feast he had laid on the table for Jesus. Then, around midnight, a struggling young mother with her child, also begging, so, without hesitation, the cobbler fed them, too. Before dawn, a boy accused of theft also visited him. Finally, after the break of dawn, Jesus visited, and the old cobbler had to admit he had nothing left to feed his savior. Jesus replied that was okay, as he had already visited three times, that very evening. As intended, the story left me with a reminder that opportunities for divine intervention happen when you expect them least.

With that parable in mind, I headed to the airport for a business trip. On the way back, due to a chance flight cancellation, I ended up getting stranded unexpectedly in Detroit, so I descended into the bowels of the airport where a long escalator takes you down to a people-mover that whisks you underground, straight to a hotel. I was the only person down there in what amounted to a big concrete cavern, riding along, completely distracted, thinking about how much my wife and I longed for a child, wondering if it was ever going to happen for us. I pondered the homily I'd heard, too, trying to find some solace in it. Up ahead, an African American street musician played *O Christmas Tree* on a saxophone. It was just the two of us down there, so I reached into my pocket to give him all the cash I had on hand, although it was just a few bucks. As the people-mover whisked me past, I dropped the money in his cup. When I did so, he looked at me with the most compassionate eyes I've ever seen in my life and said, "May you have your baby." I did a double take but was quickly whisked along, and the musician disappeared into the distance while I wondered what in the world had just happened.

Soon enough, I arrived home, where my wife became ill, so I took her to the hospital. There, they performed some tests, and we waited interminably for results. Eventually, a nurse, who knew nothing about our struggle to have a baby, arrived and briskly announced, "Well, it looks like you're a little bit pregnant!" My wife and I fell apart, laughing and crying with joy. I couldn't help but think the blessing of that street musician had been much like a blessing bestowed by one of the beggars at the cobbler's door. You never know what blessings a higher power might give you in exchange for a simple act of generosity.

The incident reminded me that I have a destiny on this Earth, and the world is supporting me in moving toward it. Thus, I've an obligation to find and fulfill it. So, while my wife carried our first child, I went into a kind of creation phase, too, pondering what kind of husband, father, friend, citizen, and Christian I wanted to be as I entered this exciting new phase of fatherhood. The rest of my life was before me, and I knew it was up to me to find my purpose in life and to leave a legacy. Finally, I wrote a purpose statement for my life: "To raise children that live a life fulfilled and to help create businesses that improve peoples' lives."

Ever since then, whenever I have a major decision to make, I check in with that vision and make sure that the decision moves me ever closer to that vision. When my children grew up, I changed the phrasing from "raise children that live a life fulfilled" to "create opportunities for children to live a life fulfilled," so that I could apply my vision to all children, not just my own. Other than that, my vision has remained the same ever since that Christmas of 1997.

Having this North Star has led to me taking on many initiatives through my employment, eventually transferring to a different employer whom I felt would help me fulfill my vision better, and taking on many volunteer projects, such as becoming a business mentor in a sustainable business accelerator. To create my own company. And now, in writing this book, I hope to disseminate the knowledge I've gained from thirty-seven years in business and product development toward humanity-centric ideals. Sure, the day might come when I retire from business, but I'll never retire from pursuing my vision. It isn't just what I do, anymore. It's who I am.

Because I'm very open and honest about my vision—at work, at home, and among my friends—you could say it has become my personal brand. A personal brand is de facto your name and your brand is what everyone says about you when you are not in the room. Whether it is discussion on your promotion or your job performance, there is a room that you are not in, and when your name comes up, your brand is on display. Typically, this reflects how much and in what way they value you. In any employment situation, every employee offers a certain amount of value. But when employees take a new job or start at a new company, they initially consume value with the office space they use, the training they require, and the salaries they make. Each of us has the job of assessing how much value we create versus how much we consume. Ideally, every employee is at, or working toward, the maximizing of value creation. The more value we create, the stronger our personal brand. As such, each of us should constantly gather data on our work to create graphs like this one:

When is Your Breakeven Point?

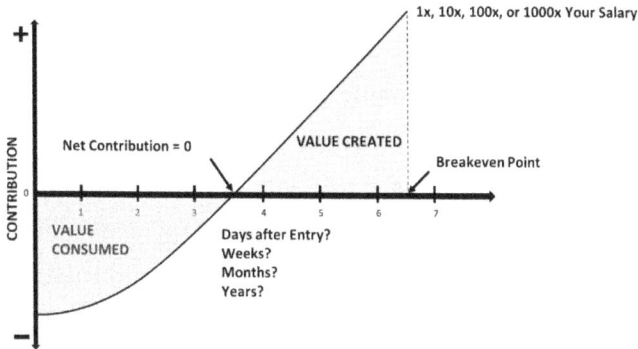

At first, each of us only consumes value, but eventually we start to create value at the same rate we consume it. This is the point of net zero contribution.

You can evaluate your own value by asking, "Did it take me days, months, or years to reach the net-zero point?" Obviously, the shorter time it takes to get there, the better. Ask yourself what you can do to get there faster.

Eventually, with continued value creation, the area under the value-created curve becomes equal to the area under the value-consumed curve. That's the breakeven point. In reality, though, no new hire will break even like that, right away. It takes time, and employers understand that, but it's important to understand what value you are creating and how to maximize it. So, as an employee or owner of your own business, it helps to constantly seek ways to minimize the value-consumed hole that you dig and accelerate your value creation to the breakeven point in order to contribute ever higher value to your company, customer, or society.

Once you have achieved the breakeven point, ask how you can maximize the value you create, ever increasing the multiple from 1x your salary to 10x to 100x to 1000x and beyond. Again, the more value you create, the stronger your personal brand. And the stronger your personal brand, the more opportunities you have to create value.

Those who create quantitative value for their clients, customers, fellow employees, and employers build trust, and the greater trust you build around you, the better your brand. Being a person who has a personal vision (and advertising it with your words, behavior, actions, and outcomes) makes you somewhat dependable, which makes you more brand-able. It enables people to know that while you'll never simply do as you're told, you'll always act in the best interest of your vision and those you serve. With a clear vision and strong brand, you have the ability to find organizations that are a good fit for you and where you are a good fit for the organization. When you have a large gap between your personal mission and the mission of the organization, this can cause stress, but where this is a good overlap, this can create passion.

If this way of looking at things works for you, you can use it to constantly increase the value you create. Each of us should evaluate ourselves and our associations based on the value we bring to others, recognizing that true worth is measured by contribution, not perception. A clear personal vision serves as a guiding principle, aligning our actions with our purpose. Mastery of this vision, rather than the pursuit of status, leads naturally to the right opportunities—whether in a dream job or self-employment—by drawing like-minded individuals to a shared mission.

Maslow's Hierarchy of Needs

No discussion of personal vision and the quest for ultimate fulfillment can really begin without mentioning Maslow's Hierarchy of Needs. It's captured below in Maslow's Expanded Hierarchy of Needs[7] that illustrates why we strive for what we do in life, in the order that we do.

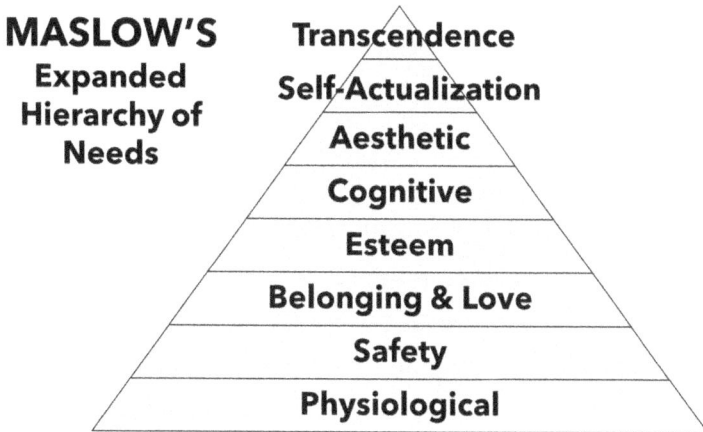

MASLOW'S Transcendence
Expanded Self-Actualization
Hierarchy of Aesthetic
Needs Cognitive
Esteem
Belonging & Love
Safety
Physiological

Maslow's Expanded Hierarchy of Needs is like a roadmap for what makes us tick, a pyramid that starts with the absolute basics and stretches all the way to something bigger than ourselves. At the foundation, it's simple—everyone needs food, water, shelter, and rest. These are non-negotiable. But in today's world, we've got to add clean air and safe drinking water to the list. Technology, too—because like it or not, access to information and connection has become essential. The need for reproduction is baked into our biology, though people experience it differently. And here's the thing: while some are struggling just to secure these fundamental needs, others, often in more fortunate circumstances, are just as stressed about challenges further up the pyramid. Perspective matters.

Once the basics are in place, the next priority is safety. That means financial security, stable employment, good health, and freedom from danger. These days, it's not just about having a roof over your head—it's also about feeling secure in the digital world, knowing your data and privacy are protected. And only after we feel safe do we start thinking about something deeper: belonging. Humans are wired for connection. We want to be part of something, whether it's family, friendships, a community, or even an online group where we feel seen and understood. You're not going to be out searching for love and commitment if you don't have a safe place to sleep at night. One thing at a time!

Then comes esteem—earning respect, feeling accomplished, and gaining confidence. We all want to be recognized for what we bring to the table, whether that's in our careers, creative pursuits, or personal achievements. But life isn't just about work and status. At some point, we crave more—learning, discovery, beauty, and creativity. That's where the cognitive and aesthetic needs kick in. Curiosity drives us to seek knowledge, and a love for beauty makes us appreciate art, music, and nature. These are the things that make life richer.

Near the top of the expanded pyramid is self-actualization—that moment when you feel like you're truly becoming the person you were meant to be. It's about pushing boundaries, setting goals, and reaching your full potential. But Maslow didn't stop there. He later realized there's something even higher: transcendence. This is where people go beyond themselves, searching for meaning in something greater—spirituality, service, or dedicating themselves to a cause bigger than their own success. It's the kind of fulfillment that lasts.

Of course, life doesn't always move in perfect order. Sometimes we backtrack. Sometimes we juggle multiple levels at once. And that's okay. The key is recognizing where we are, understanding what we need, and remembering that while our struggles may feel overwhelming, they exist within a bigger picture. Whether we're fighting for survival or searching for purpose, Maslow's Expanded Hierarchy of Needs reminds us that we're all climbing the same pyramid—just at different places on the journey.

Ikigai: True Purpose or Passion

No discussion of personal vision and goal setting for your life, or for the *rest of* your life, is complete without a mention of the Japanese concept of Ikigai.[72] Ikigai means one's true purpose or passion, and it comes from an understanding that we're all healthiest and happiest when fulfilling that very thing. Young people often get pushed into one profession or another—a wrong path leading to unhappiness—based upon what others expect. Ikigai tries to help you avoid suffering from the tendency to be a people pleaser by using a very logical process that helps you understand your own destiny. Of course, most of us have the potential to engage in many different vocations, but a Venn diagram associated with Ikigai helps us narrow down the choices.

First, we ask ourselves what we really love to do. Let's say you play a sport or an instrument or write short stories or do logic puzzles or design great outfits or make people laugh. As you answer this question, it's important to understand you don't have to be good at any of the items in this list. You just have to enjoy them. Have fun listing what you enjoy and let it be completely impractical. It's all about what brings you happiness.

With the next question, try to forget about everything you did in step one. Here, you ask yourself what the world really needs. Studying the UN Sustainable Development Goals will help with this, but you can also look at local issues. Where have you noticed in your own life, family, neighborhood, or social group that change is sorely needed? Or perhaps you're more affected by some type of suffering you've read about in the news. Then again, maybe you see and experience human suffering every day and it won't take but a minute to think of innumerable important human needs, the lack of which has led yourself or people you personally know to states of desperation.

As you make the list, don't worry about whether you know the answer to the problem. Presumably you don't know the answer, or you would have fixed it, already. So, just list the problems you're aware of locally, nationally, and internationally. It might be a long list, and that's okay. This doesn't obligate you to anything. It's just part of the Ikigai exercise.

The third circle in the Venn diagram asks what you're good at. Here, you're not to focus on what you enjoy but rather the things you've been lauded for. What subjects did you ace in school? What skills do people complement you on? If you have an established career, what are the basic skills that helped you succeed? You might think of some things you have always been naturally good at but don't particularly enjoy for one reason or another. Don't let that get you down. Just list every area where you've already seen some success.

You can list hard skills like building fences, designing software, driving long distances, researching projects, playing video games, or computational mathematics. But don't forget to also list soft skills like showing empathy, being the life of a party, caring for animals, being spontaneous, logical problem solving, and being a born leader.

Finally, the fourth circle asks that ultimate practical question: what can you get paid to do? Remember that this book is about humanity-centric innovation as an aspect of the business world, not as a volunteer activity or side hustle. If you want to make your living doing something to preserve the human race on Earth, this fourth circle helps you get down to where things get real. Luckily, in today's world, with the internet and the crowd economy providing ways to earn a living that never before existed, this list could become quite long but go ahead and let your imagination run free. Even if you're not qualified to get paid to do certain things yet, you can learn with the right effort. Determining your Ikigai is not about what you can do today with what you've already got but setting a vision for your life (or the rest of your life) in the big picture. It might take a few years' worth of education or experience to get to the point of getting paid for some of your interests, but this is not the time to worry about the tasks that are subsets of your goal. Just make a list of professions you truly think you could get paid for, or would like to get paid for, even if it requires extensive education or training.

Now, as the Venn diagram illustrates, you'll want to look at the place where these four circles overlap. Having made these lists, ask yourself if there is a skill or profession that immediately jumps to mind as something you'd be good at, enjoy, make money from, and that would do significant good in the world? For some folks, a whole list of answers imme-diately emerges. But if not, try to invent a perfect profession with your imagination.

For instance, if you:

- love to play video games,

- are good at making people laugh,

- see lead-tainted water as a big problem in your community,

- and think you could make money as a professional gamer or stand-up comedian . . .

Don't despair; just get creative. Have you heard of Al Franken, Sonny Bono, Ronald Reagan, Ukraine's Volodymyr Zelensky, or the Czech Republic's Václav Havel? They're all performers who found success in politics, trying to make the world a better place and using their humor and charming personalities to smooth their way to the top. Someone must do it. Why not you?

Then again, if you...

Love to fly fish, barbecue, and play with children

Are good at being a born leader and making business deals

See ocean pollution and world hunger as problems that desperately need solving

And already have experience buying and selling real estate but don't particularly enjoy it...

It's clear that your loves don't coincide with your established career, so try to imagine a humanity-centric profession that never existed before. With the internet, technology, and increasing ease of worldwide travel, there's a good chance you could create something that excites you. How about looking into island nations being threatened by rising tides and ocean pollution? Could you get into ecotourism, helping improve the land and water as you do so? Or perhaps your Ikigai lies in the lucrative fly-fishing industry, where wilderness fishing guides make profitable use of wild, undeveloped land and water resources. In doing so, you could contribute to the economies of local, under-served communities.

Ultimately, doing the Ikigai exercise should be fun. If it helps you draw a conclusion that was pretty obvious to begin with, that's okay if it feels like your true Ikigai, but in many cases, a lot of creativity is needed to pull these disparate elements together. I've said it before and I'll say it again: times are changing fast, and new opportunities in business are sprouting up everywhere. The time to remain in traditional fields, doing things the tried-and-true way is limited. Make yourself aware of all that modern tech makes available and enjoy the abundance of new opportunities!

The Four S-Curves of Life

In addition to finding your Ikigai, here's another way of finding your purpose and passion in life and has been a guiding light for most of mine. In my view, your purpose and passion evolve through life as you persevere through all types of adversity.

Presented here are the four S-curves of life. The y-axis represents your fulfillment and satisfaction in life, which I call purpose. The x-axis is your effort toward and commitment to that purpose, which I call passion. Each s-curve is a curve of persistence. You move through the s-curves of life through a combination of purpose, passion, and persistence.

1 Thessalonians 4:3, "This is the will of God, your sanctification."

The first s-curve is "struggle to success." No matter what you begin in life—learning algebra, playing tennis, or creating your IKIGAI—you struggle at first. But as you persevere, you eventually hit an inflection point where, with a little bit of effort, you can achieve success. Your work has become fun, easy, and profitable! You are now a person who contributes more value than you consume. Eventually, however, the gains in fulfillment level off. When the line goes flat or dips down, no matter how hard you work or how much money you make, you simply don't get as much fulfillment out of it. This can come from boredom, a sense of wanting to challenge yourself more, or a realization that money isn't everything, and you have other dreams to fulfill.

At this point, if you want to heighten your sense of fulfillment in life, you need to level up to the second S-curve, which I call "service over self to gain greater significance." This S-curve kicks in once you've hit a certain level of success, where survival is no longer the main concern. Now, you want to expend your effort in some type of service to others that gives you a feeling of greater significance in the world. The first s-curve is about adding value to yourself. The second s-curve is about adding value to others. I remember going through this phase myself. I was lucky in that my first chemical engineering job paid well and gave me a strong sense of accomplishment. But only a few years had gone by before my satisfaction leveled off, and I resolved to take stronger action toward the clean water and world sanitation issues about which I was passionate. Luckily, I informed my superiors of this vision and managed to make that transition within the same company.

For some, though, getting to the top of that initial S-curve of material success takes most of their working life. Everyone has a different path. And for many, making the subsequent transition to the second S-curve involves much bolder steps than those I had to take, but it's always worth it, as stagnation is a recipe for neither happiness nor helpfulness. But even the fulfillment you get from selfless service can stagnate after a while, which leads to the third S-curve: "surrender your ego to gain salvation." This is not only seeking your own salvation but the salvation of others, and the salvation of future generations.

Each of these curves represents your persistence in the quest toward your true purpose in life. Now, some people may decide that financial stability is enough, but in my opinion, this is limits your quest for personal fulfillment but also simply because we live in a day and age when the world needs everyone's help to sustain the human race. We, as a planet, are living in the eleventh hour, so why not get more personal fulfillment by reaching out to solve larger-scale problems and maximize that top level of Maslow's Hierarchy?

Acting in service to others surely gave you a greater sense of your significance in the world, so it fed your pride a bit, made you feel good about yourself. You needed that, then, but as your increased efforts finally ceased to bring you a greater sense of significance, it became time to stop thinking of yourself and your reputation and put others first in a bigger way. With this S-curve, the salvation you seek is not your own but that of society as a whole, now and in the future.

When you surrender your ego to seek salvation, you're going undercover, to some degree. I think my mentorship with the Toilet Board Coalition served this purpose for me, as I was able to help other entrepreneurs succeed while receiving no credit, no profit-share, no fee, nothing but a sense that I was helping the world be a better place through humanity-centric innovation. This volunteerism could be considered more charity than business work, on my part, but my efforts were expended toward helping others build businesses and led to a deep sense of fulfillment in that way. The work I did also reflected well on the company I worked for, which received accolades for sponsoring my outreach. Yet, for me, even in the quest for world salvation, my sense of fulfillment eventually leveled off.

With this last S-curve, I'm letting you into my personal world a bit. Some would argue the term "sanctification" has no place in a book about business, but this book is really about innovation for the sake of saving humanity. I feel many would agree that it takes a connection to a higher power to really comprehend doing good in an unselfish way. That's why I call the fourth S-curve: "sacrifice self to seek sanctification."

Most folks have some sense of a higher power, whether it involves an organized religion or simply an understanding that we are all part of nature's perfect ecosystem. Coming to grips with our place in that larger spiritual world or universal ecosystem brings us to a sense of fulfillment even greater than working for the salvation of mankind and planet Earth. It's working to do what's deeply *right* on a much higher level than simply what's practical.

Overall, the theme of the four S-curves of life is simply that persevering toward your own ultimate happiness is a worthwhile goal. This process isn't a straight line but a series of effortful pushes interspersed with plateaus. Our job, as strivers for satisfaction and greater purpose in life, is to recognize the plateaus when they appear and make the conscious decision to take life to that next level in our own inimitable and passionate ways.

4

WHAT INNOVATION LOOKS LIKE, TODAY

Change will not come if we wait for some other person, or if we
wait for some other time. We are the ones we've been waiting for.
We are the change that we seek.

— Barack Obama, 44th President of the United States

In this chapter, I'll draw the basic roadmap needed to achieve any innovation, not just the humanity-centric kind. First, let's define our terms. Outside the engineering world, and even inside that world, innovation is a term with myriad definitions. I describe it as the process of introducing new ideas, products, or services that create value and drive the growth of a business.

The Basic Concept of Innovation

Innovation simply means solving a problem with some new invention, idea, or system. So, consider what problem you have both the drive and the skill to solve, and remember: a true innovation must do a lot more than provide the same old thing in a new package. It must solve a problem nobody else is solving—or, as Nobel Prize winning biochemist Albert Szent-Gyorgyi put it, "Seeing what everybody has seen and thinking what nobody has thought."[73]

To design such a product, we begin with a **value proposition**. Generally speaking, in any type of product innovation, newly developed products are responses to tension existing in the minds of consumers, also called "perceived pain points." Thus, innovative engineering takes place at the creative collision of:

"What tension in customers' lives need to be resolved?"

"What are they willing to pay for?"

"What profit motive (or 'strategic ambition') does your business have?

"What solution to the tension is possible to achieve through science and technology?"

For instance, in the line of work I was in for many years, diaper design, we listened to consumer surveys to learn about the needs of moms, babies, and toddlers. When parents complained that diapers leaked, we worked harder to innovate diapers with a snug yet comfortable grip all the way around. When we learned that modern, working parents were not able to change diapers as frequently as they once could, we innovated even more absorbent diapers that could keep wetness away from babies' skin longer. When we learned that certain toddlers took longer than others to achieve toilet training, we invented a diaper that was really a type of disposable underwear, for larger children, so they could learn at their own pace. Each innovation resulted in products that addressed a pain point in the lives of babies and their parents, thus, each was highly profitable. Now, American parents have justifiably high expectations for diapers, because, over the years, innovation has met every imaginable need and then some.

Having identified a problem you'd like to solve, I suggest you go no further until you've developed a compelling vision by imagining a world changed by your product. Innovators should understand how much better this world will be, how people will change their behaviors and lifestyles because of the innovative product, and how the world at large will improve because of it. Developing a vision is more than a goal-setting exercise—it's a dream of a better world through science and technology. With your vision in place, you'll next want to run a pilot. Keeping in mind, with today's technology, this can be a virtual product in a virtual process with virtual testing, before you ever make a physical prototype or product. If the pilot works, scale up the project accordingly, and voilà, you've invented something. But you still haven't become an innovator.

To get your innovative idea off the ground, the next step is to select a business model where all the partners in the value chain receive an acceptable return on investment. In the diaper example above, certain non-woven materials must be manufactured, and a diaper factory must then convert the product. In addition, packagers print and manufacture the plastic, paper, or cardboard containers for the product. All the companies working on different aspects of the product must be compensated for their work and make a profit. Next, we've got to get that product to consumers.

In this phase, innovators develop a go-to-market strategy that determines which stores will sell the product and for how much. Discussions between the manufacturer and various retailers take place as to where the new product will be displayed on store shelves. Nowadays, this includes online sellers, as well, who will list the product on their virtual shelves, featuring it in the agreed-upon manner. In this step, advertising alerts potential consumers to the existence of the new product and the way it addresses a particular perceived pain point.

Having developed a value proposition and vision, run a pilot, selected a business model, and organized a go-to-market strategy, you can begin to manufacture and distribute your innovative product. Once consumers purchase the new product and find it solves a perceived tension in their lives, the innovative company begins to profit from sales, which further instill consumer trust in the brand and shareholder trust in the company. This is when consumers begin to see this particular company as a pipeline of innovation that will continue to solve the pain points in their lives.

By using the innovative product, consumers also develop trust in its claims. If the diapers claim to hold wetness for up to two hours without causing diaper rash, and they can actually do this, customers develop what we in business call "a reason to believe," meaning, essentially, trust in the brand. From there, the profitable and highly esteemed business can identify additional perceived pain points and innovate new products to serve those. And on and on it goes.

Humanity-centric Innovation Versus Consumer-centric Innovation

Humanity-centric innovation contrasts with the consumer-centric innovation that profit-driven companies have pursued for centuries. While one model has been around for centuries and the other is only now being introduced, they're similar in many ways. Humanity-centric innovation addresses essential needs of humanity and the planet, focusing on sustainable and societal impact. In contrast, consumer-centric innovation aligns with consumer desires to enhance company profitability, often by addressing perceived needs or 'pain points.' While both drive innovation, they differ significantly in scope, focus, and end goals.

Consumer-centric innovation happens at the creative collision of what's needed by the consumer, what's required by the business, and what's possible through science and technology. You can start with any one of those elements, but all three are required to have a successful innovation. For example, you might identify a compelling consumer need, but without a viable business model and the technology to bring it to life, it remains just an interesting observation. Likewise, you could have the greatest technological breakthrough of all time, but without a consumer who wants it and a business model to support it, it's merely an impressive, unused invention. And, while a strong business model is important, without consumer demand to drive it and technology to enable it, the concept lacks purpose. True consumer-centric innovation happens only when these three elements — consumer need, business viability, and technological capability — come together, creating solutions that are not only groundbreaking but also relevant, profitable, and achievable.

Consumer-Centric Innovation

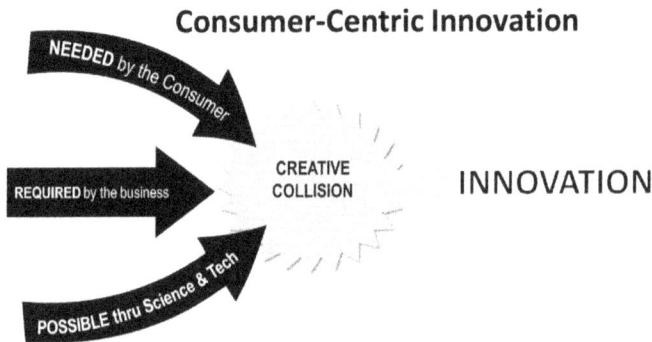

For instance, many people perceive a higher-priced brand of a certain product to be superior to a lower-priced brand of the same product. In fact, many consumers purchase those higher-priced brands because it makes them feel they are rising in social class. This is a good example of a truism in the business world: people don't buy products; they buy better versions of themselves.[74] The theory that the more expensive product is in some way superior may or may not actually be true, but many companies produce higher-priced versions of their products, which are sometimes the exact same thing in different packaging, to fulfill the needs of consumers experiencing tension around wanting to associate with being the best, purchasing the best, and generally being more high class than others. Such products do fulfill such customers' psychological needs, but only by creating an illusion. By contrast, humanity-centric innovation stays away from creating illusions for profit. Instead, this innovation model identifies real, pressing needs of the planet and works to solve not "perceived" needs but actual ones that threaten our societies and environments. Certainly, humanity-centric innovators must design products that make a profit, but these products must also make sense from the perspective of real planetary need.

Humanity-centric innovation broadens the innovation lens to include the needs of humanity at large, often aligning with the UN SDGs as a guide. These innovators must look beyond profit to consider the larger social and environmental impact of their work, aiming to contribute positively to global issues, not just avoid doing harm. To this end, modern humanity-centric innovation employs exponential technologies like AI, robotics, genomic sequencing, virtual reality, and additive manufacturing, which address critical challenges in issues like health, education, sanitation, and environmental sustainability.

Humanity-Centric Innovation

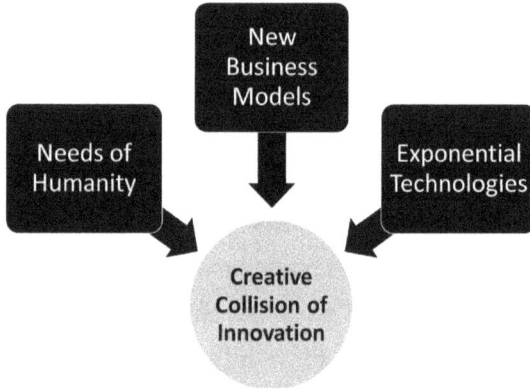

New Business Models

Needs of Humanity

Exponential Technologies

Creative Collision of Innovation

This type of innovation has the goal of benefiting society at large, including social, environmental, and economic systems. It emphasizes ethical considerations along with sustainability challenges and social inclusivity to improve the quality of life for a specific population. Essentially, humanity-centric innovation contributes to the greater good while also meeting consumer needs; however, doing so within a profit-generating company requires a deeper level of thought than old-fashioned business did. That said, another difference between humanity-centric and consumer-centric innovation comes into stark relief when we look at the process of goal setting,

Differences in Goal Setting and Competition

Within any company, setting goals for innovation begins with truly understanding that firm's core competencies. When I was leading humanity-centric innovation in one of the companies where I worked on superabsorbent products, I wanted to know the essence of our work, so I asked experts what they thought our company's core competencies were. Confidently, we believed these were absorbency, fibers, and non-woven fabrics. While this was true, it didn't tell the whole story. As we looked deeper, discovering that the products we create are for vulnerable people. Our customers never ask us for "non-woven fibers." They ask us for ways to meet their essential needs and make their lives better.

When parents come home with their first baby, they're entering a whole new world where they have lost control of their lives. When a twelve-year-old girl has her period for the first time, she's lost control of her life, too. If an older person has incontinence and sneezes at a party, they've just lost control, too. In each of these cases, we discovered how our company provides a product that restores lost control and dignity. This is especially important to those who lose control in taboo situations having to do with bodily functions, where it's a lot harder to ask for help.

For instance, when teams developed the first successful line of adult diapers, the engineers were not motivated by talking about nonwoven fibers and superabsorbency but by imagining a world where one could live seventy, eighty, ninety, or even a hundred years without the last five years being filled with the depression, loneliness, sadness, and pain that statistics show is nearly inevitable. We asked, *What can we mere engineers, with our limited skill set, do to help these people who are, after all, our own parents and grandparents?*

With that in mind, we were indeed finally able to realize our company's core competency: "To empathetically provide control and dignity to people in socially sensitive areas such as menstruation, incontinence, diapering, and toilet use." The non-woven fabrics and absorbent products we produced were simply a means to that end. Understanding your company's core competency establishes your overarching goal, although each individual product you innovate will have its own sub-goal, as well. This type of deep goal setting is typical of social-change-oriented, humanity-centric innovation.

Another big difference between the consumer-centric and humanity-centric innovation models resides in the area of competition. Consumer-centric products all typically have their own competitive landscape, where opposing companies offer something similar but with different qualities, features, or price points. In that case, finding the right price point for your product is a big part of the challenge of achieving market domination. But humanity-centric innovators don't look at it the same way. They often provide products nobody else offers to solve problems where competition does not exist. If other providers do exist, humanity-centric innovators should ask themselves if their own product is still necessary or how they can partner more effectively. In a humanity-centric approach, the focus shifts from simply 'beating the competition' to driving meaningful progress. Companies innovating for sustainability often find that collaboration, rather than pure competition, leads to greater impact and long-term success.

Don't Limit Yourself to What's Possible

The potential of humanity-centric innovation is to bring every individual on Earth to the same standard of living as the wealthiest one percent, resulting in a lifestyle (which is quite common in developed nations, already) that is so luxurious it surpasses that of royalty a mere two hundred years ago. Modern technology, infrastructure, communications, and healthcare feature such world-changing innovations as vaccines, electricity, indoor plumbing, clean water, and instant world-wide live video calls. Even with their armies of servants and elaborate horse-drawn carriages, royal lifestyles of old could never approach this now-ordinary level of comfort and convenience.

Technology has not only raised our lifestyle expectations but made it ever-more possible to ensure that raised standard becomes evenly distributed across the planet. As such, humanity-centric innovation occurs at the intersection where sustainable development goals meet innovative business models powered by exponential technology, driving significant, positive change all over the globe. If this still sounds idealistic, you would do well to realize that today's technology requires you to fundamentally change your baseline notion of what's both possible and affordable, as those metrics are constantly changing. As is often attributed to Nelson Mandela, "It always seems impossible until it's done."

Speaking of what's possible, note that in today's tech landscape, it's important not to have a fixed mindset or an old-fashioned way of thinking. For instance, many would-be innovators limit their imaginations to what's *currently* possible, but with technology developing as fast as it is, there is a good chance some new, competing tech will pop up while your business is still in its infancy. There is also a chance technology exists that you just don't know about, yet.

Take, for instance, what's going on in space tech right now. While space programs don't have a colony on Mars yet, our equipment has been there and even recorded a Martian sunrise.[75] We have equipment in space that studies Mars quakes, recording data that tells us just what a year on Mars is like, day by day. In other science news, the Large Hadron Collider at CERN (Conseil Européen pour la Recherche Nucléaire, or European Council for Nuclear Research) has used particle accelerator technology to achieve temperatures hotter than the sun with high energy particle beams traveling at nearly the speed of light.[76] We're also conducting infrared astronomy with the James Webb telescope, which conveys images from its position in outer space. There, this incredible technology uses high-resolution and high-sensitivity instruments to capture images too old, distant, and faint for even the Hubble telescope to detect. This amount of knowledge about the distant universe is unparallelled by anything humanity has achieved before.

What's more, scientists have captured an image of a black hole by using radio telescopes that "shake hands."[77] They work in pairs to contribute small pieces of information gradually compiled by AI to create a new visual understanding of what is really going on in outer space. While the image of a lone astronomer peering through a mountaintop telescope remains iconic, modern astronomy has evolved into a highly collaborative field, where advanced technology and coordinated scientific teams work together to capture and analyze images of space.Whether you're as fascinated by space technology as I am or more concerned with terrestrial pursuits, these groundbreaking advances must give you pause. I mention them here to make you realize we are living in unprecedented, phenomenal times. And you're a part of it. How much a part of human advancement you'd like to be is, of course, entirely up to you.

I observed a great example of how fast our progress is really zooming forward when I saw two magazine covers from *Forbes*[78] and *Time Magazine*[79] published on the exact same day: November 12, 2007. The *Forbes* cover lauded Nokia for achieving one billion customers and asked, "Can anyone catch the cell phone king?" Meanwhile, *Time Magazine* introduced the iPhone as one of the best inventions of the year.

Nokia's undoing, in fact, was already underway! This kind of technological leapfrogging is happening right now in innumerable different industries, but rather than let the speed of our world overwhelm us, it's incumbent upon us denizens of this constantly evolving dynamic to select a niche where we can thrive and do our best to ensure the progress that's made is made toward sustainable goals. That's all each of us, in our own small lives, can do. And—like radio telescopes sharing tiny bits of information with one another until they create a single complex picture—our individual efforts really do add up to something far greater than the sum of its parts.

Remember that in 2007, when those magazines were published and humanity thought itself on the brink of a cell phone revolution, there was still no Instagram, no Uber, no Bitcoin, no WhatsApp, no Airbnb, no Tesla, no 4G connectivity, no blockchain technology. Almost twenty years later, those technologies have a market cap greater than two trillion dollars and new, world-changing technologies never stop getting invented. Perhaps you yourself will be part of a team pioneering the next revolutionary phenomenon!

I think I've duly made the point that technological progress is tumbling forward, gaining momentum like a snowball rumbling down a mountainside. With that in mind, you've really just three essential questions to ask yourself to find your niche in this exciting, dynamic, never-boring world.

> *What's impossible today but, if it could be done, would fundamentally change my business or chosen profession?*

> *What's impossible today but, if it could be done, would fundamentally change society for the better?*

> *What's impossible today but, if it could be done, would fundamentally change my life, forever?*

Being on the cutting edge of innovation means constant awareness of emerging trends, repeating patterns, the speed of change, and the key players in your world to let those clues direct your actions. It takes a lot of practice to maintain such a high level of hypervigilance, but trust me: do it enough, and it becomes intuitive. Add to that an awareness of the seventeen UN Sustainable Development Goals (SDGs) and an inner drive to make positive, long-lasting change for a billion people, and a humanity-centric innovator is born.

In a world where innovators build a better "mousetrap" on a regular basis, you must always have a view to what's possible rather than what currently exists. Ironically, this means that if you are a person who was ever reprimanded for being a dreamer, you're just the kind of person the world needs, today. In fact, when I see the word "impossible," I like to add an apostrophe and a space, so it reads, "I'm possible."

Zipline's Unique Innovation

A company called Zipline is a great example of innovation that takes advantage of high-tech robotics to do some truly humanity-centric work. Its founders saw a problem that needed a solution, where nobody else had stepped in to help: when medical workers in remote areas of the world ran out of supplies, they became all but useless to the individuals they were supposed to be helping.[80] Often located in war-torn or famine-affected areas, clinics had to close due to their inability to obtain supplies in a timely manner, and lives were lost as a result. But Zipline has harnessed drone technology to deliver supplies to just such hard-to-reach regions. As such, this company has been able to transform healthcare logistics, specifically in Rwanda and Ghana, by providing lifesaving medications and blood products without risky journeys. What's more, Zipline is no charity. It's a profitable company, which means it won't run out of funds, be canceled by a shift in government funding, or expire for lack of volunteers—such is the benefit of the humanity-centric business model.

The Trend Away from Atoms and Toward Bytes

One central tenet of the movement toward humanity-centric innovation is the fact that companies are trending away from selling atoms (physical things) and toward selling bytes (digital products). The more we as a society can digitize our many products and manufacturing processes, the fewer of the earth's resources we will use to produce them. We'll also have fewer waste and recyclable products to dispose of when the item in question reaches the end of its lifespan. However, the energy demands of digital infrastructure, such as data centers and IT systems, mean we must approach digitization thoughtfully to ensure it contributes to sustainability all around. A great example of this phenomenon is the book you're reading right now.

Some readers prefer the tactile sensation of a paper book, and hardbacks and softcovers still exist to fulfill that preference, but eBooks and audiobooks fulfill many readers' needs with less waste, no deforestation, no need for shipping, and no waiting time. Within the past decade, the market for digital books and magazines has grown significantly, while print book sales have fluctuated but remain a substantial part of the industry. Even when buying paper books, though, many people do so via the digital process of online shopping. Such is the power of bytes over atoms.

Of course, we're not in *The Matrix*. We do still live in a world filled with actual things made from real resources. However, digitization has also become part of the development of atom-centric products like cars, clothes, guitars, and so many other things because today's innovation often begins by creating digital prototypes based upon existing atom-centric products, perfecting those products in the digital realm, then creating new and improved versions of the atom-centric product. It's more than asking, how do we go from atoms to bytes, but rather how do we use these data to make better products and provides our customer with new experiences.

For instance, imagine if you wanted to build your dream house. First, you would go looking at real, existing houses similar to what you want. You may like the open-plan living area in one, the infinity pool in another, and the bedroom suite in a third. Your visits to all those existing, atom-centric homes inform your idea of what you want to build. Next, an architect would take your design ideas and create a virtual prototype of your dream house on a computer. This way you can see the house, as it emerges from the architect's imagination, from all sides and even from above. You can see cut-away views and even digitally design your landscaping while you're at it. You might even be able to use virtual reality to go inside the prototype of the house and feel what it's like to walk around in there!

Once you and the architect have perfected the digital prototype, builders will construct the actual, atom-centric home. If this technology didn't exist, you would still have blueprints and sketches to refer to, but these are more abstract representations of the house. You'd have to use real resources to build the house before being able to know what it would actually look and feel like. If, after the house was built, you realized you didn't like it, a lot of money and resources would be wasted in rebuilding! Nowadays, this kind of digital prototyping exists in nearly every atom-centric innovation, so no matter what products your company manufactures, if you want to keep improving them over time, affordably and efficiently, a humanity-centric approach will guide you to save resources by entering the digital realm.

Beyond digital prototyping, 3D printing is now revolutionizing the way homes are built. Instead of relying solely on traditional construction methods that require significant time, labor, and materials, large-scale 3D printers can rapidly create entire structures layer by layer using concrete, bioplastics, or other composite materials. This approach dramatically reduces waste, speeds up construction, and lowers costs—making homebuilding more efficient and sustainable. One of the largest projects demonstrating this shift is the 100-home community being built in Austin, Texas, using 3D-printed concrete structures. A collaboration between ICON, Lennar, and BIG-Bjarke Ingels Group, this project showcases how automated construction can provide high-quality, affordable housing at scale.[81] These homes are not only built faster than traditional methods but also designed to be highly energy-efficient and resilient. The combination of digital design, virtual prototyping, and 3D printing is reshaping how we think about architecture and construction, making it possible to go from imagination to reality with unprecedented efficiency.

History has shown the march of progress from atoms to bytes through the rise and fall of companies well known to most adults today. As mentioned above, Apple's cell phones, with their intuitive digital interface, replaced the formerly dominant mobile phone provider Nokia, back in 2014. Blockbuster, founded as a video rental shop, used to be America's favorite place to get tapes and DVDs for home players, but when Netflix's digital offerings came along, they took over the market. Now, people of my generation laugh at the bad old days when we had to remember to rewind VHS movies lest we be charged a penalty upon return. Kodak, once the dominant company for all things photographic, disappeared once digital cameras and social media apps came on the scene, and celluloid film is as outdated as burning CDs for a playlist. Similarly, ride-sharing apps like Uber, with its all-digital interface and lack of a centralized physical hub, has, in many places, made the act of hailing a cab on a city street a distant memory. Although, in many major cities, hailing cabs still remains a key part of urban transportation, with many people still relying on traditional taxis alongside ride-sharing services.

Today, both consumer-centric and humanity-centric innovators are going digital, both to save on costs and resources and simply to serve the needs of an ever-digitizing population. Add to that the fact that technologies like blockchain, AI, and virtual, mixed, and augmented reality are constantly improving the digital landscape. Unlike in the physical world, the possibilities for digital innovation are basically limitless. Humanity-centric innovators should rejoice at the possibilities presented in this regard, most especially because the developing world is becoming digitized, too—a fact that will eventually allow innovators to solve many social and environmental problems there, far more easily. These efforts are helped by innovations like vastly improved batteries for long-term energy storage, which will put more rural areas on the digital map. Meanwhile, advanced robotics enable atom-centric work to be done in regions where people can't go (like natural disaster areas and outer space and the insides of human bodies) while people safely, accurately, and digitally control them from afar.

5

NECESSARY TECHNOLOGY
FOR INNOVATION

You look at science as some sort of demoralizing invention of
man, something apart from real life, and which must be cautiously
guarded and kept separate from everyday existence. But science
and everyday life cannot and should not be separated.

— Rosalind Franklin, a British chemist whose work was central to
understanding human DNA

In our rapidly changing world, it is simply impossible to truly innovate without using the most up-to-date, exponential technology relevant to your business. For companies founded on the latest tech (and those that update to it) you'll find every aspect of both business itself and the process of achieving sustainability much easier and less expensive. As such, you and your team must have established a digital IQ around generative AI based upon data standardization. That's why I'm talking primarily to Gen Z, because many of you grew up immersed in exponential technologies, naturally understanding and appreciating the potential to be entirely humanity-centric. But only if we choose to use them that way.

Artificial Intelligence

The AI Universe

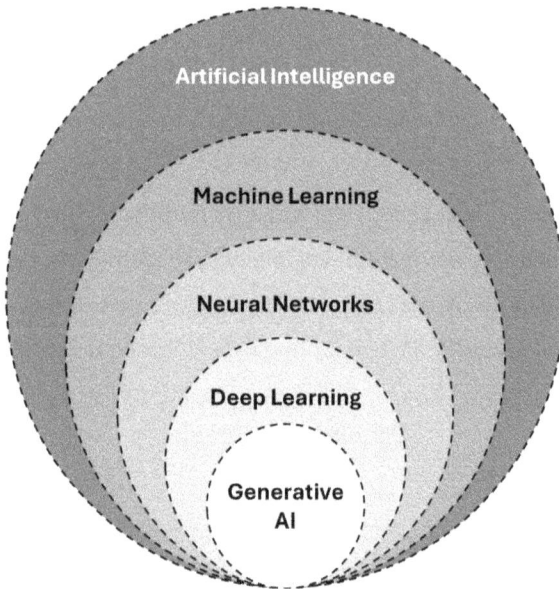

Artificial Intelligence

Machine Learning

Neural Networks

Deep Learning

Generative AI

Back in the late 1800s, during the second industrial revolution, if you wanted a good idea for a new business, all you needed was to take an existing tool, say a drill or a washboard, and add electricity to it, thus creating a power drill or a washing machine. Electricity was the magic business bullet of its day. Today, that magic bullet is AI. For instance, cell phones have become "smart phones," washing machines are now "smart appliances," and cars are well on their way to becoming autonomous vehicles. Yesterday, cars had computers. Today, cars are computers on wheels. Tomorrow, all cars will drive themselves as opposed to just some of them. The AI world is the "smart" world, but AI won't just build profitable businesses, it truly has the power to help save the planet for the human race. To use it effectively, though, innovators must not just utilize but deeply understand today's artificial intelligence landscape.

Success with products featuring AI relies upon innovators constantly seeking to build upon their own digital IQ, including elements such as deep learning and machine learning, generative AI, and neural networks. While all modern tech is helpful for humanity-centric innovators, artificial intelligence, in particular, is crucial because we humans are in a hurry to solve planetary problems before the greenhouse effect makes the planet unlivable for humans. So, at this point in our evolution, we are truly in a race against time to save ourselves. Luckily, AI does whatever analysis and decision-making people used to do, only much, much, much (have I added enough "muches" to convince you?) faster.

For instance, imagine what the world will be like once self-driving cars have been perfected. Whenever you drive, you make thousands, maybe millions, of decisions in the blink of an eye. You pay attention to the gas gauge, the speedometer, the rear-view mirrors, the traffic around you, and the road conditions ahead to make life-saving adjustments. If anything is off in drivers' concentration, such as being in a highly emotional state, they can make mistakes with far-reaching consequences. But with an AI-driven vehicle, a machine can process data and make decisions significantly faster than a human could, relying on algorithms designed for optimal problem solving without emotional biases influencing the decision-making process. You'll get where you're going faster and more safely while being freed up to focus on your personal mission statement. Traditional humanitarian work can sometimes feel like driving through fog—slow and uncertain. But with AI, the road ahead becomes clear, accelerating our ability to create meaningful, lasting change with greater speed and impact.

Machine Learning, Deep Learning, and Generative AI

One type of AI, called machine learning, works by recognizing patterns—but first, it needs a little human help. Imagine you're building an adoption website for a local animal shelter, and your program needs to sort images of dogs and cats. At first, a human has to teach the AI what's what: dogs usually have prominent noses and floppy ears, while cats often have whiskers and pointed ears. The AI studies these patterns, learning to tell them apart. Once trained, it can sort new images on its own, saving tons of time and effort. This is how AI powers everything from recommendation algorithms to self- driving cars—it learns from data and gets smarter over time. Deep learning takes AI to the next level by cutting out the need for human guidance. Instead of being manually taught what a dog or cat looks like, deep learning AI trains itself by analyzing massive amounts of data. It processes millions of images using neural networks—an advanced system designed to mimic how the human brain learns. Over time, it gets smarter on its own, recognizing patterns without any human input. The result? AI that can accurately categorize millions of images in seconds, powering everything from facial recognition to self-driving cars. Generative AI is a subset of deep learning AI, which can not only identify and categorize things and make decisions based upon what it "learned," but also generate its own images and text by analyzing the data provided. For instance, ChatGPT generates unique text based upon textual data provided to it, employing what it learns from the data regarding content, grammar, style, and vocabulary.

When guided by responsible leadership, all these AI protocols make it much easier than ever before to develop humanity-centric solutions. And, over time, AI-enabled solutions will be more effective than ever because they'll be profoundly scalable, cost-effective, and have a global reach. What's more, AI enables people all over the world to communicate and collaborate on digital platforms, making mass collaboration and decentralization nearly effortless. It's a profound irony that, in our quest to unite humanity across borders, cultures, and beliefs, we may find our best ally in artificial intelligence. But we must remain vigilant, for AI's path forward could either strengthen human solidarity or undermine our autonomy. Make no mistake about it, the speed of AI computing is the key to unraveling this planet's dangerous imbalances. We need this speed. With the accelerating pace of climate change, nuclear tensions, and the growing threat of pandemics, no other problem-solving method will ever be fast enough to fix them. For certain tasks, deep learning offers a speed advantage over traditional machine learning, but its need for massive amounts of data brings in a paradigm-shifting topic I'll address in just a few pages: why data is now so coveted, it's a new type of currency that can change the world.

As computing power continues its exponential growth, businesses that seem impossible today—like fully autonomous AI-driven companies, real-time brain-computer interfaces, and personalized medicine tailored at the genetic level—will soon become viable, unlocking entirely new industries and transforming the way we live and work. This assertion, in a way, has been reinforced by the greatest hockey player that ever lived. When asked what made him such an incredible player, Wayne Gretsky is known to have quipped: "It's simple. Don't skate to where the puck is. Skate to where it will be." Gretsky wasn't psychic, he just knew how to identify familiar patterns and spot trends on the ice and therefore predict the future just a little bit. Experience doing so enabled him to anticipate an opportunity that didn't exist when he was standing still. We innovators must do the same.

Three Different AI-Focused Business Models

AI can now analyze years' worth of data in minutes to draw conclusions it would take mere mortals or ordinary computers weeks or months to achieve. AI is a data-hungry monster, and as such it can take just about any data a company collects and draw conclusions, like how to automate low-value tasks or eliminate bottlenecks in production. AI's process optimization method is more efficient and makes fewer errors than manual processes. Its consistency is unmatched, and its performance is so reliable that it's become an affordable, low-risk way for businesses to stay in a state of continuous improvement. It's like having a tireless, always-learning system that keeps evolving to make everything run smoother and smarter. That's fine for the business world, but what of the planet? Indeed, AI can also be used to analyze data related to climate change, pollution, and world conflict in order to offer data-driven solutions. But different types of AI strategies are needed for the three primary manifestations of AI used in business: AI optimized, AI enabled, and AI fueled.[82]

AI Optimized

At its most basic level, process optimization is a simple matter of constantly analyzing your performance to improve the way you do things. Anyone can use it. For instance, a piano player could analyze her playing to quantify the number of times she gets her timing wrong, then get a better metronome or a different practice schedule or a new teacher, then re-analyze her performance to perceive any change that took place. If her playing improved, her process optimization worked well. Similarly, businesses have long used process optimization to increase efficiency, effectiveness, and quality towards a better bottom line. In fact, when you receive a survey about the quality of the customer service you received, that's part of a process optimization effort toward constant de-bugging and improvement of that system. But now, with the introduction of AI, process optimization has leveled up with more speed, accuracy, and efficiency.

In the true spirit of process optimization, AI-optimized businesses insert AI algorithms into their automated processes so that optimization is always going on in the background. American Express is a good example of this, as its recently launched AI-optimized system provides enhanced fraud detection ensuring AI instantly notifies all parties if fraud is detected by the system.[83] While this crucial work is automated, the human-run aspect of the system partners with legal, privacy, compliance, procurement, and cybersecurity experts to round out the security protocol. Meanwhile, the customer service related to helping potential identity-theft victims need not be automated. In such a stressful situation, a human touch is clearly needed, thus enabling machines to do what they do best while keeping people involved in the humanity-centric aspect of things.

AI Enabled

By contrast, AI-enabled business models provide optional AI automation, enabling consumers to choose automation rather than making it a required part of the system. Tesla is famous for making best use of AI enabling with its vehicles' optional self-driving features. Drivers always have the option of operating their own vehicles—after all, driving a high-performance sports car is a thrill! But when owners would rather put things on autopilot, they can upload software that makes the car go faster, manages the battery life, and performs all other aspects of getting from one place to another. In this case, AI optimization records and analyzes drivers' use of the self driving feature in order to improve it.

Coursera is another well-known AI-enabled company. It offers the option of using AI to personalize learning experiences and tailor course recommendations to individual students' needs. This AI-driven approach improves student engagement and success rates and democratizes access to education, providing additional growth opportunities for students of all ages.

AI Fueled

AI-fueled businesses are often also called "AI first" companies. In this model, everything is automated. During the COVID 19 pandemic, Moderna Pharmaceuticals provided an excellent example of this type of technology at work. After

Chinese scientists published the genetic sequence of SARS-CoV-2 on January 10, 2020, Moderna's AI-powered, in-silico vaccine-development system quickly went to work. Within just three days, by January 13, Moderna had finalized the design for its mRNA-1273 vaccine candidate,[84] marking one of the fastest vaccine design processes in history. By February 7, 2020, Moderna had already produced the first clinical batch of mRNA-1273, and by March 16, it entered Phase 1 clinical trials.[85] This lightning-fast timeline marked a historic breakthrough, demonstrating how technology, collaboration, and science could come together to confront a global health crisis in record time.

AI also managed drug design, manufacturing optimization, and even patient recruitment for the vaccine's clinical trials. A year later, many lives were saved when citizens all over the world had the COVID vaccine injected into their arms. Process optimization within the AI-fueled system ensured not only efficient vaccine development but the same system's ability to improve itself and provide better vaccines periodically, over time.

Babylon Health also uses an AI-fueled system to provide affordable, accessible healthcare for a wider swath of the population.[86] With telemedicine offerings that include virtual consultations, symptom checks, and health monitoring, as well as AI-powered diagnostic machinery, this medical company drives positive social change by reaching out to those unable to travel to hospitals or clinics for a variety of reasons.

The distinctions between AI-optimized, AI-enabled, and AI-fueled technologies matter because innovators must consciously choose the right approach to AI in order to manage risk, resource allocation, and the skill sets of their employees and technologies while optimizing processes. The type of AI a company utilizes also has a lot to do with the level of innovation and disruption to which it aspires as well as helping accurately represent any new product's capabilities. These terms also help ensure transparent communication about any product's use of AI, which helps consumers (and investors) set realistic expectations and build trust. However, for startup founders interested in this technology, it's important to know that AI-first companies often face increasing regulatory scrutiny, particularly around privacy, data security, and ethical considerations.[87]

Virtual Realities

Virtual reality and its variations have emerged alongside AI as crucial aspects of modern technology that not only transform the way we interact with the world but also offer many potential humanity-centric benefits. Variations of these digital realities include Virtual Reality (VR), Augmented Reality (AR), and Mixed Reality (MR), all of which can promote enhanced learning, improved healthcare, and sustainable practices.

Virtual Reality, the most well-known of the digital "realities," utilizes headsets such as the Oculus Rift, HTC Vive, and PlayStation VR to immerse users in a virtual environment, isolating them from the real world. Users see and hear computer-generated environments where motion-tracking sensors follow user movements and controllers enable them to interact with the fully immersive virtual environment. As you may already know, this technology relies on high-resolution displays and spatial audio to create a sense of presence so convincing that, after using the technology, users often experience the equivalent of "sea legs," where it takes some time to re-adjust to Earth-bound reality.

VR isn't just for entertainment. It's also used for educational experiences, scientific exploration, and to improve healthcare, such as with the Osso VR, which trains surgeons by simulating complex surgical procedures that enable medical students to practice and refine their skills in a risk-free environment[88]—an innovation that supports UN SDG #3 (Good Health and Well-being) by ensuring better training for healthcare providers.

Augmented Reality, on the other hand, utilizes the same basic technology to overlay digital information onto the real world for the purpose of enhancing user perception through devices such as smartphones, digital tablets, and AR glasses. This effect is created using cameras and sensors that capture the real world, then add computer-generated images, sounds, motion tracking, environmental understanding, light estimation, and other data on the video.

For instance, VirtualOn's Virtual Fashion Fitting Mirror enables users to stand in front of an augmented-reality mirror and virtually try on clothes.[89] The system provides automatic, accurate body measurements in about three seconds, along with fast browsing and selection of clothes, which customers can easily "change" into and out of. The system even enables users to move, see how the fabric moves with the body, and view themselves from all angles. Then, if desired, a simple click enables online purchases—essentially turning your closet into an unmanned department store. This futuristic waste-and-fuel-reduction idea enables basic commerce, but on a higher level, AR can also be effectively utilized to promote sustainability in earthbound reality. For example, Pale Blue Dot is an AR app that raises awareness about climate change by overlaying visual data on real-world locations, showing the impact of environmental changes so that users can see potential future scenarios based on current trends, which supports UN SDG #13 (Climate Action).

AR has also shown itself to be especially instrumental in preserving resources and preventing waste, specifically by streamlining retail operations to reduce shopping time, materials, and fuels used to transport goods. Nowhere is this more important than with bulky, heavy purchases such as furniture. To this end, IKEA Place is an AR app that allows users to visualize IKEA furniture in their homes before making purchases.[90] By providing a realistic preview of how products will look and fit in their space, IKEA's AR enhances the customer shopping experience and reduces the likelihood of returns, promoting responsible consumption and supporting UN SDG #12 (Responsible Consumption and Production).

Mixed Reality technology combines elements of AR and VR, allowing enhanced reality and virtual elements to interact in real time. It uses advanced sensors, optics, and computing power to merge the real and virtual worlds by overlaying holograms onto the physical environment, where they provide a seamless blend of digital and physical worlds. Devices such as Microsoft's HoloLens and Magic Leap are examples of current tech that enable these experiences.

In the education field, a company called zSpace utilizes MR to create immersive learning experiences in classrooms.[91] Students can interact with 3D models—such as dissecting a virtual frog or exploring ancient ruins—to enhance their understanding and retention of complex subjects. This approach aligns with UN SDG #4 (Quality Education) by providing engaging and effective learning tools.

Overall, businesses leveraging the power of AR, VR, and MR are setting the stage for a future where technology and human well-being are harmoniously integrated, paving the way for a more sustainable and equitable world. In fact, because of these digital realities, some of us are already engaging in something called Multiple World Models, which means we no longer live in only one place but have real-world personae and online personae, too. As time goes by, this type of delocalized existence is only going to expand. With the rise of Augmented Reality and Virtual Reality, humanity is introducing more layers to the equation of basic existence. Business innovators would be smart to keep in mind that with all the different possible versions of ourselves, opportunities exist for new business innovation to serve both our Earthbound and virtual selves.

For instance, Second Life, the very first virtual world,

created in 2003, gave rise to a multimillion-dollar economy where people were paying other people to design digital clothes and digital houses for their digital avatars.[92] Truth is, every time we add a new layer to the digital strata, we're also adding an entire economy built upon that layer, which leads to conducting business in multiple worlds at once. With my emphasis on humanity-centric innovation, though, I can't endorse businesses serving and receiving payments from avatars unless, of course, innovators find a way to utilize these alternative digital selves toward the UN SDGs. But who knows, maybe Multiple World Models are exactly where we need to be to innovate for a clean future. Could you be the first humanity-centric MWM innovator?

Blockchain Technology

In the quest for innovative computing solutions that align with the principles of transparency, inclusivity, and sustainability, blockchain technology stands out as one of the crucial transformative tools at our disposal. Initially recognized for its role in cryptocurrency, blockchain is now being leveraged across a wide variety of economic sectors to support the UN SDGs.

You don't have to understand blockchain to have it integrated into your business model, but I'll sum up the essence of this technology here, providing a little something for the highly computer savvy along with enough simplicity for the less technically minded. Essentially, blockchain is a decentralized digital ledger technology that records transactions across multiple computers in a way that makes data extremely difficult to alter retroactively. Each transaction is recorded in a "block," and these blocks are linked in a "chain," hence the name. This structure creates a secure and transparent way to record transactions, making it ideal for applications requiring data integrity. After all, any attempt to sell or trade data must be based upon that data's reliability. (I'll talk about the importance of consumer control over the sale of personal data a few chapters down the line.) So, blockchain is the technology that keeps your data secret when you want it to be but also enables you to ensure its reliability should you wish to sell it.

The innovative nature of blockchain technology is best explained by contrasting it with what it replaces. You see, the traditional databases computers used to use for data storage were managed by a central authority, putting all the power of data management in that authority. But blockchain operates on a peer-to-peer network where each participant has a copy of the entire blockchain. Contrary to the ideas of early computing models, users have found that decentralization actually enhances security and transparency. This is because once a block of information is added to the chain, it cannot be altered without changing all subsequent blocks, which requires consensus from the network majority, and that would be a rare event, indeed. So, blockchain-stored data is basically immutable. What's more, blockchain networks also use consensus mechanisms such as Proof of Work (PoW) and Proof of Stake (PoS) to ensure all participants in the network agree on the validity of transactions. This way, it is basically impossible to transmit false or doctored information through blockchain technology.

A good example of one of the uses of blockchain technology comes from both Stellar and BitPesa, platforms that facilitate low-cost, cross-border payments. Financial services must, above all, be secure, which has always presented difficulties when it comes to international transactions, but with blockchain ensuring security, Stellar and BitPesa are able to promote worldwide economic inclusion to people in developing regions—a truly humanity-centric endeavor.

What's more, because of its security, blockchain cuts down on greenwashing. It can verify and incentivize eco-friendly actions, track carbon credits, manage renewable energy distribution, and promote responsible consumption—all by ensuring the accuracy of data. For instance, a company called Power Ledger uses blockchain to create decentralized energy markets promoting renewable energy sources.[93] Veridium is another example of a sustainability-oriented use of blockchain.[94] It uses the technology to manage carbon credit transactions to ensure the integrity of carbon-offset initiatives.

Blockchain also ensures the worldwide supply chain of food by accurately tracking the journey of food products and providing unassailable data that ensures ethical sourcing, chain of custody, and reduced waste. Companies such as IBM Food Trust,[95] Provenance,[96] and others have revolutionized efforts toward the eradication of hunger by relying on blockchain technology. Blockchain's transparent, inclusive, and sustainable solution to these and many other global challenges shows that sustainability is not a retrograde idea. Today, moving forward with technology is the only realistic way to further humanity-centric goals. In fact, blockchain enhances sustainable practices so well that it has become a truly transformative tool for global development. But tech advances are only one aspect of humanity-centric innovation. To utilize them for sustainable goals, innovators need to understand the new economies in which they can be utilized.

6

THE NEW ECONOMIES: INNOVATION OF THE FUTURE

Someone's sitting in the shade today because someone
planted a tree a long time ago.

— Warren Buffett, investor and philanthropist

Most people in the business world are familiar with the typical consumer-centric economic model where products are sold directly to the consumer in exchange for money. It's a simple, reliable type of transaction, and we rarely stop to think there might be some other way. Meanwhile, if someone gets something "for free," it is understood that someone else, perhaps the government or a charity, are "paying for it." In this case, three parties (the buyer, the seller, and the user) are involved, but it's still a direct transaction where the buyer pays the seller for the product before giving it away to the user. However, other economic models exist. Those most affected by climate change, often in developing countries, are rarely the ones with the greatest influence over its causes. Wouldn't it be transformative if humanity-centric innovators embraced new economic models that empower these communities, ensuring they have the resources to build resilience and adapt, rather than bearing the financial burden of addressing global environmental challenges? One of the unique economic models innovators are looking at these days is called "circular," because it emphasizes the idea that the economy does not need to be a straight line from producer to consumer. The reuse and recycling of household products have long been common examples of circular innovation, which emphasized saving both money and resources by reducing waste. It also acknowledged the fact that raw materials could be processed and utilized more than once. This basic idea, now known as the "cyclical economy" is now just the most primitive version of a circular economy, as it lacks an emphasis on big-picture, industrial change. Whenever you recycle a paper bag back into new paper products, you're engaging in cyclical thinking. This approach, part of the circular economy, aims to keep materials in use by reprocessing them instead of discarding them, ultimately reducing resource extraction and waste. It's a great place to start thinking and acting in a circular-economic way. But what we currently call "the circular economy" serves as an umbrella concept for many other alternative economies that enable active trade in a variety of non-linear ways.

The Closed-Loop Economy (Industrial Symbiosis)

The closed-loop economy is a much more far-reaching concept than simple recycling and represents the type of big-picture thinking this planet needs for real change. A closed-loop economy imitates nature, where the waste of one species becomes the food of another. Here, every industrial activity is analyzed for the waste it produces then paired with another that harvests that waste for use as fuel or feedstock

One popular example of this model in action is called "regenerative agriculture—" today's answer to the original idea of organic farming. Using this technique, fallow land is planted with certain crops such as clover or fava beans that will not be harvested but tilled right back into the earth to sequester nitrogen in the soil. When a profitable crop is grown in that field the following season, the nitrogen that was a waste product of the first crop becomes free fertilizer for the next one.

Another more-industrial example of this model in action is Denmark's Kalundborg Symbiosis, a pioneering version of a consciously designed, industrial, circular economy.[97] There, the Asnaes Power Station generates electricity, producing steam as a byproduct.[98] Danish pharmaceutical giant Novo Nordisk then uses that steam to run its own production systems.[99] In turn, Novo Nordisk's organic waste and bio-mass is sent to local farms for fertilizer and biogas plants for energy production. Meanwhile, Gyproc, a plasterboard manufacturer, also uses steam from the power station for its own manufacturing, which produces gypsum waste, which is recycled via donation to a cement manufacturing company.

Within the system, all significant waste is turned into a different type of fuel.

South Korea's Ulsan Industrial Park is another example of a complex network of industries, each feeding one another with their waste. In this system, an oil refinery produces excess heat, hydrogen, and sulfur. The sulfur is piped to nearby chemical manufacturers, and the Hyundai motor company uses the heat and hydrogen in its fuel cell factory. That factory, in turn, produces scrap metal, which is then sent to a company called POSCO Steel for recycling. The steel company, in turn, produces slag and dust, which are trucked to a cement manufacturer to be integrated into their product. Overall, the entire linked system reduces waste and improves efficiency.[100]

The Netherlands' Port of Rotterdam is yet another example of a closed loop economy operating in a truly circular way. There, Shell refinery produces CO_2, but instead of being released into the atmosphere, it's piped to local greenhouses to enhance plant growth. Those greenhouses, in turn, produce organic waste, which is recycled into energy by a local bio-fuel plant, which, in turn, supplies fuel right back to Shell refinery, completing a fully circular industrial ecosystem.[101]

Loop: Pioneering the Circular Economy for Sustainable Households

In the quest to transform our economic model from a linear "take, make, dispose" system to a circular one, a company called Loop has emerged as a groundbreaking initiative.

Operating on the small scale to serve households, Loop specializes in innovation for waste reduction. Developed by TerraCycle, Loop promotes sustainability in its own way, aligning closely with several UN SDGs.

Loop is fundamentally designed to address the inefficiencies and environmental harm caused by the traditional linear economy. Its platform is focused on creating durable, attractive, reusable packaging that can be cleaned and refilled, thereby eliminating the need for single-use containers. This system not only conserves resources but also significantly reduces waste, creating a more sustainable consumption model.

Loop's unique packaging system promotes sustainability in three ways. First, it is conveniently standardized in size and shape for efficient collection, cleaning, and reuse. Second, the Loop system promotes consumer convenience with its subscription and delivery service for collecting and reusing containers, which integrates seamlessly into modern lifestyles. Third, the company's partnerships with major brands such as Unilever, Procter & Gamble, Nestlé, and PepsiCo ensure a wide range of popular products will be able to utilize the system. For instance, Unilever's Dove and Axe products are available in Loop's stainless-steel containers,[102] as are Procter & Gamble's Tide detergent.[103] Nestlé's Häagen Dazs ice cream is available in a specialized double-walled steel container,[104] and PepsiCo's Tropicana orange juice is offered in Loop-approved glass bottles.[105]

Overall, Loop's circular economic model not only addresses environmental challenges but also creates economic opportunities and enhances social well-being. The company exemplifies the transition modern businesses must make from a linear to a circular economy. In fact, considering all the technology currently available to create seamless and efficient systems, one can't help but wonder why every atom-based business isn't on board with some type of circular-use packaging system.

With Loop's systemic approach to waste reduction, it promotes UN SDG #12 (Responsible Consumption and Production) by providing long-term consumer savings. It also bolsters UN SDG #11 (Sustainable Cities and Communities) by fostering a culture of responsibility and circular product use while also generating jobs in recycling, cleaning, and logistics, which contribute to economic growth and community development. Finally, UN SDG #3 (Good Health and Well-Being) is a focus of Loop's system as well, for by reducing plastic waste and pollution, Loop contributes to healthier environments and addresses health risks associated with pollution. Overall, Loop demonstrates that sustainable practices can be both economically viable and socially beneficial, driving progress towards a more sustainable and equitable future.

The Blue Economy

The "blue economy" is a term coined by economist Gunter Pauli[106] to describe an economic model that emphasizes sustainable and innovative use of natural resources, inspired by ecosystems. While it includes the sustainable use of oceans and seas, it goes beyond that, aiming for resource efficiency and zero waste across industries. The blue economy seeks to drive economic, environmental, and social change by transforming waste into resources and fostering carbon-neutral, self-sustaining solutions. The blue economy is no pie-in-the-sky idea, as many companies have already found profitable ways to benefit from zero waste across industries. Fish waste recycling in aquaculture is a powerful example of the blue economy, creating a zero-waste, circular system in aquatic farming. In this model, nutrient-rich waste from fish farms is repurposed to grow seaweed or shellfish within the same ecosystem. Seaweed, for instance, absorbs excess nutrients, improving water quality while producing an additional crop that can be harvested for food, biofuel, or animal feed. Meanwhile, filter-feeding shellfish, like mussels or oysters, further purify the water as they grow, providing a sustainable source of seafood. This model not only reduces environmental impact but also creates new revenue streams, combining profitability with real sustainability for a cleaner, more efficient way to farm the ocean. The current profusion of offshore wind farms and coastal energy projects are great examples of the way the blue economy is being taken more and more seriously as an economically viable example of a circular economy. These energy projects promote UN SDG #7 (affordable and clean energy) by generating renewable energy that reduces reliance on fossil fuels and lowers carbon emissions while also generating jobs and fostering clean-energy tech advancements. The reliable access to clean energy generated by much of the blue-economy further supports social and economic development all over the world.

One emphasis of the blue economy is the fact that plastic pollution in the ocean is not only unsightly and dangerous, but it also breaks down into micro-and nano-plastics found in all living organisms, including humans. Nobody quite knows the long-term effects these nonbiodegrading pollutants will have on human health, but we do know the best way to reduce the growing problem is to simply nip it in the bud by eliminating oceanic plastic waste altogether, addressing both UN SDG #12 (responsible consumption and production) and UN SDG #14 (life below water). Innumerable companies formed around coastal cleanup are now serving sustainably minded communities and thriving as they do so.

The Plastic Bank

The Plastic Bank is a great example of a closed-loop economy using blockchain technology both to eradicate poverty and clean up the environment, all at once.[107] With its simple but revolutionary idea, the company confronts the crucial issue of financial exclusion, where individuals and societies known as "the unbanked" have no access to bank accounts or the ability to borrow money. At the same time, The Plastic Bank also confronts the different but related issue of poverty, where individuals' incomes don't provide for life's essentials.

This type of company, known as a FinTech, uses up-to-date technology to improve the financial standing of underserved communities. The simple idea is that individuals who gather plastic garbage of all kinds and deliver it to Plastic Bank locations can be paid for it via a digital wallet within the Plasticbank app. In many cases, this is the first bank account such users have ever had. Plastic is subsequently recycled, which offsets the cost of the program to some degree, but most of the funding for this for-profit business comes from partners, including small and large businesses as well as individuals, who sponsor The Plastic Bank.

This subscription-style sponsorship helps reduce the participants' environmental footprints and promote a circular economy. For companies wishing to promote their social action, The Plastic Bank's blockchain-verified subscription model provides transparency, allowing them to share their sustainability efforts with stakeholders and customers. Studies show that 88 percent of consumers seek to support brands that help them live a more eco-friendly lifestyle, so a relatively inexpensive "impact subscription" to The Plastic Bank appeals to customers' values which then impacts the company's reputation, ultimately ramping up market share.

The Plastic Bank's positive impact upon the environment can not be overstated. We now know that plastic trash entering the ocean breaks down into micro-and nano-plastics that invade our bodies and significantly affect sea life at all levels, yet no technology has yet been developed to effectively remove it. Thus, The Plastic Bank does the only thing that truly can be done at this point, which is to prevent plastic pollution from entering oceans, rivers, forests, deserts, and urban spaces alike, working toward environmental change from a preventative standpoint.

The Plastic Bank takes pride in the fact that it is different from companies that sponsor one-off beach-cleaning
endeavors or build machinery designed to gather plastic
directly from the ocean. Instead, it stops plastic waste at
the source by engaging local underserved communities all
over the world in regular, ongoing, profitable, and enjoyable
plastic-gathering activities in which the whole family can
participate. Meanwhile, The Plastic Bank's blockchain technology ensures subscribers that their money is being well
spent, the plastic is being effectively recycled, and communities are being served exactly as expected.

Water purification is another type of blue-economy
innovation that's not as simple as it may, at first, appear,
but a company called IDE Technologies has pioneered
water recycling to create potable water for drought-stricken
communities, including desalination plants.[108] Addressing
UN SDG #6 (clean water and sanitation), this company's
advanced tech provides clean water with minimal energy
wastage. Its technique for ensuring a reliable potable water
supply also reduces operating costs, all while improving
public health and economic stability in water-scarce regions.

While many blue-economy projects are already up and
running, numerous others, like revolutionary algae-based
biofuels, are still in development. Scientists have recently
discovered that algae stores a great deal of oil, which can be
used to make fuels similar to our current carbon-based fuels.
Corn is already being used in this way to make ethanol, but
algae could be a game changer because its oil content can be
as high at 80 percent as opposed to corn's 5 percent.[109] Also,

because algae absorbs carbon dioxide, it can be grown in industrial areas to clean the air while potentially replacing the fossil fuels causing pollution in the first place. Best yet, algae's oil can be extracted via sound waves, and it doubles in size every twenty-four hours.[110] I could go on. There are innumerable benefits to innovating with algae, but currently the manufacturing process requires more energy than the final product can produce, so this is still a fuel-of-the-future. That said, could your innovation team be the one to bring it across the finish line?

Innovators in the blue economy discover and refine these types of hidden natural gems, which will one day provide the type of self-sustaining renewable fuel that will, in theory, provide all the necessities of life for free, via the gifts already existing in nature.

The Decentralized Autonomous Organization (DAO) Economy

In many ways, today's AI provides an unexpected opportunity to truly democratize society. As such, decentralized autonomous organizations (DAOs) reshape business structures with the futuristic ideal of robot-run organizations. Built upon blockchain technology, smart contracts, AI, cryptocurrency, and VR or AR tech, these organizations can function with little to no human input or oversight and free up people to function at a higher level. That means no more "workers" and "bosses." Instead, people all over the world collaborate on the philosophical and theoretical level, then program the DAO to perform relevant tasks. The human participants in these economies are equals, with equal voting power. With DAO economies, the ultra-modern idea of having a business that manages itself dovetails nicely with humans doing what they still do better than computers—use imagination to innovate unique solutions in sustainability.

On the developer side, economies based upon DAOs democratize investing by enabling participants to pool resources and expertise to invest in online business. They reshape business hierarchies with an inclusive and democratic ethos. They also foster transparent systems: since everything is automatic and self-executing, the rules of the DAO are universal. It's an idea that could change governance forever. In terms of global application, DAO economies are, for the most part, still in the theoretical stage, as this type of big-picture thinking requires a clear international tax framework and non-conflicting international legislation, but these are the ideas behind an innovative, automated, tech-driven, humanity-centric future.

DAOs are most well known for their use as gaming platforms. For instance, Ready Player DAO's gaming software is designed to harness and develop the collective powers of play-to-earn gaming on a platform featuring decentralized and permissionless metaverse games.[111] But DAOs have also revolutionized fundraising for sustainable projects. Klima DAO is a great example.

Klima DAO asserts that climate change is the result of an international coordination failure due to the fact that aligning sustainability incentives globally is difficult, largely because of the opacity of the way the Voluntary Carbon Market works and the resulting lack of trust.[112] Klima DAO works toward a solution by having autonomous protocols that function as a market maker for environmental commodities within a system made transparent by blockchain technology. Participants can become token holders and voters, submit and review proposals on a form, and play a part by applying to be part of the review committee.

SoCity DAO

DAOs are uniquely poised to promote prosocial, pro-sustainability behaviors in cities, too. SoCity DAO, for instance, was initiated in 2020 by students and researchers at the City Science group, MIT Media Lab, and the City Science Lab @ Shanghai, who focus on using technology to improve urban design and generally improve city life. Theirs is an ecosystem that serves to quantify each individual or organization's social or sustainable value contribution according to its behavior, and then monetize and recognize that contribution to reward users and further incentivize the positive feedback loop. What's more, the SoCity DAO platform empowers people all around the world to start their own SoCity initiatives by providing resources like smart contract toolkits for economics, governance, privacy-preserving data, data analytics, funds, advisories, and community networks. Naturally, blockchain, smart contracts, and AI data analytics ensure a secure, efficient, and equitable process.

The Crowd Economy

This interesting economic paradigm shift is already in full swing all over the world, and it's quickly changing social structures. Here, an economy is based upon generating revenue from existing, under-leveraged assets. This is where your Airbnb space-sharing, your Uber and Lyft ride-sharing, and like-minded businesses fit in. These types of businesses have scaled at speed by shifting entire regions from product-based economies to service economies with no new physical infrastructure.

These models also succeed by rejecting the old notion of "full time" or "part time" employees (along with the office infrastructure that used to define a business) and leaning, instead, on the modern staff-on-demand model, which is agile enough to adapt to a rapidly changing environment. Amazon's Mechanical Turk, which provides web developers with virtual access to micro-task temp workers, exists on the low end of the digital crowd-sourcing movement, while Kaggle's data-scientist-on-demand services excels at the high end.

Crowd economy services are also a good example of the modern trend toward selling "experiences." Here, uniqueness is valued. The predictable same-ness that used to attract travelers to hotel chains has fallen into disfavor as people enjoy the crowd economy's new opportunity for adventure and slight risk.

Economists say the crowd economy "maximizes existing assets, increasing productivity with greater efficiency," and that it leads to "more consumption, enhancing economic growth."[113] All true, but on a more emotional level, it humanizes travel and transportation with endless, affordable variety. What's more, the crowd-sourcing economy extends into more deeply sustainable ventures such as Hello Tractor, an app that connects African and South Asian farmers with local tractor owners willing to rent out their equipment on an as-needed basis.[114] This convenience reduces overhead for farmers, enhances agricultural productivity in regions where it's needed most, and replaces a competitive atmosphere with a sense of community.

Kiva's Crowd Funding for Microloans

A company called Kiva is a great example of how automation and the concept of crowd funding can profoundly change the lives of hard-working entrepreneurs and solopreneurs, even in the most far-flung regions.[115] Run by investment professionals, this company crowdfunds microloans for unbanked people all over the world such as refugees, displaced people, communities impacted by climate change, and systemically marginalized people here in the US. Eighty percent of the funds tend to go to women, and the data shows Kiva's work therefore makes a great impact on building long-standing gender equity.

Typically, Kiva loan recipients are hard-working individuals who simply lack access to credit, yet their credit needs are quite modest. For instance, a mere $100 loan could purchase machine parts or chickens or seeds or other materials that could truly transform a small business such as a farm, delivery service, or laundry—all within a legitimate business relationship that eschews charity. The crowdfunding aspect of the business ensures that investors—not banks but ordinary people just like you and me—can contribute whatever amount feels comfortable and the Kiva system will aggregate all the contributions until the recipient receives the requested funding. Over time and bit by bit, the loan recipients pay back the money, which is deposited right back into each lender's Kiva account. Kiva's 94.6 percent average repayment rate demonstrates that both recipients and loaners consistently show commitment to the program. Once lenders' monies are returned, they can retrieve the sum from their accounts or simply loan it again to another recipient, enabling an amount as low as $25 to travel the world, doing good for everyone it touches.

Kiva's system depends upon automated processes that disburse and receive funds according to pre-set protocols—utilizing AI or an AI-like interface. The company also partners with local microfinance institutions that accept the loan applications and lending partners who enable the necessary wire transfers, so collaboration and decentralization are also paramount aspects of this business, making it fully modern, cost-effective, and global in its reach, however Kiva is a 501(c)(3) California nonprofit. As I've stated, my ideal model of such a humanity-centric business would be one with a profit motive. Kiva currently receives two-thirds of its funding from donations and grants, which is great as long as the money keeps flowing, but I think in order to grow and eventually serve a billion needy people, the world should spawn more

Kiva-style microlenders who have calculated into their highly automated systems some way for the company itself to profit.

The Experience Economy

Where business innovation used to provide either unique products or unique services, it now tends to focus more on "experiences." Starbucks is a good example of this. While it does sell a product (coffee) and provide a service (café seating), Starbucks also fulfills a much deeper "experiential" need in communities—it gives people a place to hang out for customers that is reliably upscale. Thus, the coffee franchise is neither work nor home but a "third place" in which to live your life: a caffeinated theme park of sorts.

Unique, independent cafés exist all over the world, but you never know what you'll get when you patronize each of them, what the atmosphere will be, or what sort of impression you'll make on the world by being a patron. But with Starbucks, you always know. All over the world, your presence there gives an impression, a *feeling*, of worldly sophistication and quality. More than the coffee, it's that experience that's at the core of Starbucks' success—a truly consumer-centric phenomenon.

Peloton

Peloton is another good example of successful "experience" marketing. Whereas exercise bikes used to be sold for their physical features, this bike is sold not for what it is, or even for how it can benefit you, but rather for the many experiences it offers such as live classes, on-demand classes, and, importantly, a member lounge, where participants can digitally befriend one another. The bike's digital interface even provides opportunities to ride with famous cyclists and receive pro tips only available to members. Despite its physical representation as a stationary bike, Peloton's main feature is its app, which also offers strength training, yoga, boxing, rowing, meditation, and barre. For afficionados of this product, Peloton is more than an experience, but a lifestyle oriented around physical and mental fitness, which provides the *feeling* of ultimate health. That focus on feeling is the essence of the experience economy.

Experience-oriented products have changed the landscape of modern business because they generate feelings such as belonging, empowerment, sexiness, fitness, moral righteousness—and the list goes on. Nowadays, though, people are connecting more and more with the need to feel useful and like a change-maker in an increasingly troubled world. This is an excellent direction for humanity-centric, experience-oriented innovation.

The Transformation Economy

This economy capitalizes upon the natural human desire for change, variety, and growth. Building upon the idea behind the experience economy, the transformation economy relies upon consumers purchasing products as catalysts toward personal change. For instance, gym memberships could be sold as opportunities for fun exercise with friends (an experience), but in the transformation economy, they are sold as paths to personal transformation. Consumers, after all, are simply buying the opportunity to sweat in a gym, but what motivates them to do so is the opportunity to look better, be healthier, and have more energy, which could lead to better outcomes in their romance, work, and social lives. They see the purchase as a chance to transform from the person they are to the person they'd like to become.

Festivals like Burning Man and retreats featuring yoga, ayahuasca, or other non-traditional healing modalities capitalize upon the same phenomenon. Spiritual pilgrimages, guided backpacking tours, wilderness survival courses, and psychotherapy are all examples of experiential investments people make *not* for the feeling they get when participating but because they want the transformative benefits promised. Some theorize that because our high-tech world has become so convenient, it has reduced the level of daily-life challenge we used to experience, along with the transformation that naturally accompanies struggle. Perhaps for this reason, modern people seek, and are willing to pay for, opportunities they believe will lead to healthy personal growth and transformation.

Technologies employing virtual reality have made great use of the transformation economy. In fact, with so many companies engaging remote workers, many companies suffer from a lack of cohesiveness and friendliness among their employees. Where once the company water cooler or coffee pot provided a gathering spot for idle conversation, that friendly element is now gone, and its loss can bring down team cohesiveness and morale. Confetti offers transformation opportunities for such companies with its virtual reality team-building experiences. Its offerings enable coworkers to bond, even if they're working remotely, by engaging them in games such as trivia contests, escape rooms, guessing games and confidence-building exercises.

TransTech

TransTech offers both hardware and software to support transformations in the field of mental and emotional health.[116] Using the latest scientific research in behavioral change, AI, digital medicine, sleep technology, and robotics, TransTech attempts to enable the individual process of expanding consciousness, making individuals aware of their old and new views to integrate the change into a new self definition.

7

DATA-DRIVEN ECONOMIES

The most difficult thing is the decision to act. The rest
is merely tenacity. The fears are paper tigers. You can do
anything you decide to do. You can act to change and control
your life and the procedure. The process is its own reward.

— Amelia Earhart, aviation pioneer

Data is the lifeblood of the modern economy, powering
everything from targeted advertising to predictive analytics.
Companies like Google, Meta, and Amazon have built empires
on the collection and analysis of user data, translating insights
into enormous economic value. At its base, the data economy
or so-called free economy, seduces customers with enticing
free services and makes money off the data gathered from the
recipients without permission. Originally, this was a type of

bait-and-hook sales technique, but by the time the general population realized what was happening, society had become so data-driven, and consumers so addicted to free stuff, that nothing really changed, and the Big Data Revolution was on. In fact, consumers interested in selling their own products or services soon realized they, too, could benefit from access to data-based micro-demographics through online and social media pay-per-click advertising. That's how many dorm-room startups became global superpowers—a youth-empowering phenomenon with no foreseeable end in sight.

According to a report by McKinsey, artificial intelligence is expected to contribute up to $13 trillion to global economic output by 2030.[117] In such a digitized world, data has become so valuable it is being called "the new oil." And yet, just as in most of the major social paradigm shifts in recorded history, this change has resulted in power being concentrated in the hands of the few. In the past, citizens have successfully counteracted such change by establishing trade unions and cooperative banking institutions, which ultimately led to anti-trust laws, labor rights, and banking reform. Creating a balance of power has always begun with such citizen organizations, which is why today's data-driven economies hold incredible potential for empowering citizens. But their existence relies entirely upon society's ability to develop methods for individuals to control and sell their own data.

Due to the new trend in seeing data as a type of currency, humanity-centric innovators today are asking themselves how they might provide necessary recyclable products to underserved communities in exchange for nothing but data. That data would then be provided to a third-party, which would pay the initial provider for it, completing an economic circle.

For instance, let's say you're a professional acrobat. If I were to give you the latest leotard model for free, asking only that you leave a review of it online, I'd be providing an item in exchange for data (in this case, your opinion). That data would benefit the item's manufacturer by enhancing its online presence, meanwhile the manufacturer would actually pay *me*, the merchant, for providing you with this free item. In this circular model, the manufacturer values data equally or more-so than actual money.

People produce data all the time. In today's world, where DNA analysis yields such a wealth of information that it can solve decades-old crimes, even a fingernail clipping contains valuable data. Human behavior is data. Shoppers' preferences are data. Teenagers' taste in music is data. Body height, weight, and measurements are data. Peoples' social and political affiliations are data. Peoples' health statistics and dietary histories are valuable data to medical researchers. Even human excretions like blood, urine, and feces provide immense amounts of data about individuals, societies, cultures, and their health. Deep-learning AI can use much of that data to create valuable new medicines and vaccines, predict the consequences of certain actions or policies, and even learn about current trends in order to either encourage or halt them depending upon their preference. With enough data at its disposal, today's ultra-fast deep learning programs can essentially reveal the keys to human health and happiness. So, if data isn't a valuable currency, I don't know what is. Luckily, everyone has a lot of it, and, unlike gold, its value is not attached to scarcity.

Because every person on Earth is a gold mine of data, the data-driven economy has the potential to eliminate social stratification and empower the marginalized. This all depends, however, upon people not having their data stolen, so data-sovereignty is currently a big focus of innovators who understand this emerging economy. Bitcoin, blockchain, decentralized data marketplaces, and data cooperatives are all potentially helpful technologies in this regard.

For the existing data economy to become a truly circular one, critical changes in design and digital infrastructure are necessary. For instance, to ensure circularity, data-generating products must be redesigned to be durable and repairable while also being easy to break down into recyclable components. But to ensure products are data-driven in a way that serves consumers, these products (and the systems of which they are a part) must also be redesigned to collect and privately hold data until consumers choose to disburse it. Finally, for that to be effective, methods must be in place for consumers to monetize their data in order to sell it back to the product's maker or to another data consumer.

To some degree, data-driven, circular-economy products do currently exist. Good examples include cell phones, smart home appliances, electric vehicles, and wearable technology, all of which are designed to collect and transmit data to their manufacturers while also being easy to disassemble for eventual recycling. However, while these products rely on the recycling and re-use of valuable consumer-generated data, consumers themselves do not currently receive direct compensation or benefits for their contributions. A true circular economy would benefit tech users more directly than that. In such a scenario, consumers would be empowered by new or existing regulatory agencies that develop a way to explicitly govern data ownership, sharing, and monetization. After all, data trading is currently in the Wild West stage, where anything goes, and theft and piracy are all part of a game that favors corporate data pirates. Alongside regulation, consumer education to increase digital literacy would be crucial for a true circular data economy, perhaps even as a required subject in schools. In fact, the success of consumer-empowering data marketplaces and cooperatives relies entirely upon a level of education in this subject that fosters a new culture of informed data usage.

The Inclusive Data Economy

Studies have shown that when underserved communities are identified and their data collected, the data economy can more easily reach and serve their needs, resulting in more economic production and wealth growth both for the community in question and those who provide products for them. This is called "inclusive data," as it draws data from multiple sources, both traditional and non-traditional, to identify markers such as gender, social status, family structure, time use, and financial means, all in order to serve the neediest, who may not already be actively involved with sharing their data on the internet. Inclusive data collection enables the launching of micro-loan programs and other such financial initiatives that benefit both providers and users.

In the case of inclusive data (as opposed to free data, below) the data has not typically been volunteered through internet usage, so gathering it must be done ethically, with respect and confidentiality, and in such a way as to prevent replication of the biases and discrimination that may already be holding such populations back. Inclusive data is not an end in itself, but a means to informed decision-making for shaping policies that uplift underserved communities.

The Free Data or "Smart" Economy

Ever since the dawn of the internet, people have gotten used to getting data of all kinds for free, to the point where, now, people often resent being asked to pay for information that, thirty years ago, would have required in-depth library research. Whenever we use a search engine like Google or Yahoo, and ChatGPT, we are getting information, ostensibly for free. But the truth is, we are unknowingly engaging in barter. We actually provide our own data in exchange for the search engine's data. Thus, the search engine learns our preferences and shopping habits in order to direct customized advertising our way. Thus, when you search the internet, the internet searches you. Typically, we enjoy this exchange because the advertised products are often things we want to purchase.

The free data economy has changed our society in many ways, not all of them good. Consider the fact that social media apps are free data economies that collect even more personal information than internet search engines, including our family photos, likes and dislikes, and lists of our friends and associates. This information enables algorithms to customize the content sent to each of us. So, while social media provides endless entertainment, it has also taken vulnerable individuals down false-information rabbit holes and even started cults. Data are powerful, even dangerous. Nonetheless, pretty much the whole world has fully invested in the free data economy, and it is up to us to ensure that, going forward, it's used to build a positive, sustainable future instead of falling into the hands of bad actors.

Getting Data Back Under our Control

Most of the alternative economies discussed in the previous chapter exemplify the circular economy ideal, where productivity and wealth growth are part of complex relationships between entities that manufacture, purchase, sell, and use products and services. As such, simplistic, linear, buy/sell transactions are becoming obsolete. With the world changing as fast as it is, it's important that everyone from scientific innovators to household shoppers understand that how we conduct trade today can either uplift communities toward a sustainable future or drag them down, economically. For those running existing businesses or planning to start one, choosing sustainability with circular economic models is a great way to make sure you're on board with a humanity-centric future. But for circular, data-based economies to create a true economic revolution, we've got to reclaim our data from those currently snatching it for free. Luckily, numerous organizations are working seriously on this.

Data Regulation and Governance Tech

Technologies for data governance do exist, though they're little known. For instance, the General Data Protection Regulation (GDPR) is a comprehensive data protection law in the European Union that regulates how personal data is collected, processed, and stored.[118] It grants individuals greater control over their data and imposes strict obligations on organizations to protect privacy. GDPR aims to enhance data security and transparency, ensuring that personal data is handled responsibly.

Meanwhile, in the United States, The California Consumer Privacy Act (CCPA) is a state-wide data privacy law that grants California residents new rights regarding their personal information.[119] It allows consumers to know what data is being collected, request deletion of their data, and opt out of data sales. The CCPA aims to enhance consumer privacy and promote transparency in data-handling practices.

To help organizations comply with the few global regulations that do exist, OneTrust has emerged as a privacy, security, and data governance platform. It provides tools for consent management, data mapping, and privacy impact assessments, ensuring that businesses handle personal data responsibly. OneTrust supports compliance with regulations like GDPR and CCPA, enhancing data privacy and protection.

TrustArc is another privacy management software designed to help businesses comply with data protection regulations while providing solutions for managing consumer consent, conducting privacy assessments, and ensuring regulatory compliance. Importantly, TrustArc enables organizations to build trust with customers by demonstrating a commitment to data privacy and security.

Typically, data is traded without the subject's knowledge or consent. If you've ever had your credit checked, you were made starkly aware of exactly how much of your personal, financial, and work-history data is gathered by others. If you've ever been solicited for votes or donations to political parties, you've been made aware of the fact that your political leanings, ethnicity, education level, income, and other demographic data is out there, outside of your control... and sometimes it isn't even correct. For the most part, society has accepted this reality, but as awareness of the value of such data grows, an international movement of data privacy advocates is working to reverse this situation and return ethics to the world of data dissemination. There is even a movement afoot to establish the privacy of one's personal data as a human right. This path can be paved by cooperative organizations designed to rebalance the digital economy for a more inclusive and equitable digital future. After all, a rising tide lifts all boats, not just yachts.

Four billion people on this Earth are living on less than $2.50 per day.[120] Financial analysts refer to them as the "Bottom of the Pyramid" (BoP),[121] but such individuals generate vast amounts of data including mobile phone usage, social media interactions, transaction records, and biometric data, which (as of the writing of this book) remain an overlooked resource. Data marketplaces and cooperatives are the very key to producing substantial income streams and inverting the traditional wealth pyramid through the monetization of the sheer volume of this personal data. It's simple, really: as data becomes an increasingly critical asset, those who control large datasets will hold significant economic power—but only if they learn to harness it.

Data Marketplaces

Data marketplaces empower consumers to record, encrypt, and sell their data. As such, they rely upon a blockchain-based, decentralized marketplace that empowers Internet users to control and monetize personal data by pooling it to collectively negotiate better terms. This is, however, a somewhat complex endeavor, as the data you may potential sell needs to be of higher quality than what ecommerce companies can already get very cheaply from existing data ecosystems.

The central problem in developing data marketplaces is the fact that, historically, data theft yields better results than data purchase, especially when the data in question regards internet search and purchase history. Attempts at buying such data tend to rely upon consumer surveys, but, whether deliberately or by accident, consumers often fail to accurately report their internet usage. So, providing the complete sets of data companies find valuable requires a technology that enables users to record, encrypt, and provide for sale the full range of their search and purchase history without having to remember it and with a guarantee that they can't edit it. Enter: blockchain technology.

Prior to blockchain, internet-user shopping history was recorded by each individual ecommerce company. Companies could sell the data they gathered in this way, but these data purchases did not include a full picture of each customer's overall shopping habits, trends, and shifts— just the searches and purchases made per store. A more comprehensive picture is necessary for AI to effectively target consumers for future sales. Blockchain technology records individual consumer behavior across the entire internet, in chronological order.

Developing blockchain-enabled, decentralized data marketplaces-of-the-future is crucial to empowering consumers to sell their own data while ensuring it is of high quality. Such "Decentralized, Privacy-preserving Data Marketplaces" (dPDMs) do away with any centralized authority or data repository while enabling users to modify the rights over their data at will. Meanwhile, notaries audit transactions to ensure authenticity within the blockchain, prompting data buyers to utilize a querying system to request purchase data. An automated data-pricing mechanism assigns prices and performs the transactions, using a token-based system to make the trades. Tokens can be exchanged within the marketplace for gift cards, public transportation vouchers, and other benefits.

Existing dPDMs include companies such as Wibson,

Ocean Protocol, DcentAI, Datum, Dataeum, Opiria, and Streamr, but this is a new and wide-open field that still invites a great deal of innovation toward humanity-centric goals. See appendix A of this book for a list of useful data marketplaces. In addition to data marketplaces, consumers in the new data economy can be further empowered by data cooperatives.

Data Cooperatives

By contrast to data marketplaces, data cooperatives are groups of allied individuals who pool and protect their data with specific software, both to analyze and therefore improve the economic, health, and social conditions of the group as a whole and/or engage in group negotiations for the sale of that data. As an innovation within the original cash economy, credit unions long-ago originated for this same purpose, so it's possible that, as society evolves, group data management and distribution may become a secondary purpose for these very same institutions. However, specific cooperatives are also emerging within the data economy in order to leverage the power of, for instance, one hundred million US consumers.

For example, Salus Coop and MIDATA are both citizen-driven cooperatives that manage the health data of groups for research purposes. They also control sales of specific aspects of that data to approved companies and institutions. But data cooperatives don't just manage health data. In fact, LBRY, Polypoly, Eva Coop, Ubiquitous Commons, Mnémotix, Digi.me, DataWallet, Data Union, and dOrg.tech are all examples of democratically centralized consumer data cooperatives designed to ensure their communities retain control over a wide range of personal data from health-related, to environmental, to transport and mobility, to energy and consumption, to digital search and consumption patterns. (See Appendix A for a list of Data Cooperatives)

Data-Encryption Technology

Within the ecosystem of companies concerned with data privacy, Brave Browser has emerged as a necessary but little-known innovation. This privacy-focused web browser blocks ads and trackers by default, providing a faster and more secure browsing experience. Brave Browser rewards users with Basic Attention Tokens (BAT) for opting into privacy-respecting ads, enabling them to monetize their browsing data. The browser also features a built-in cryptocurrency wallet for managing BAT earnings.

Other alternative data-management technologies include disk encryption software. Encryption software enhances data security by protecting stored information from unauthorized access. For instance, VeraCrypt is an open-source disk encryption tool that allows users to create encrypted volumes and partitions. It is commonly used to secure sensitive data on personal computers and external drives. Similarly, Bit-Locker is a disk encryption feature integrated into Microsoft Windows, designed to encrypt entire volumes and restrict access to authorized users. As encryption technologies evolve, solutions like these continue to play a critical role in data protection across various devices and platforms.

Existing Innovations that Monetize Data

Above, I've focused on technology that empowers consumers to encrypt, protect, and sell their data through marketplaces and cooperatives, but to sell it, there has to be a business willing to buy. Luckily, technologies also exist for fair-trade data-purchasing. Datumize is a good example. Specializing in the capture and processing of data that is not typically accessible, Datumize enables organizations to monetize their data by transforming it into actionable insights—all to provide valuable information for business intelligence and analytics.

The decentralized Ethereum platform pairs well with a company like Datumize to enable the creation and execution of smart contracts. These self-executing contracts run on the blockchain, ensuring transparency and security. Ethereum supports various applications including those that allow users to monetize their data by establishing trustless agreements between data providers and consumers.

Utopia Analytics is another company that provides tools to analyze and monetize data, but it specializes in artificial intelligence and text analytics. This company's solutions enable businesses to extract valuable insights from unstructured data and create licensing agreements for data usage while supporting data-driven decision-making and enhancing the value of data assets.

How to Monetize your Personal Data

As the saying goes, "a journey of a thousand miles begins with a single step." So, if you're interested in innovating around the data economy, it's a good idea to start with yourself. Here's a simple yet comprehensive guide to help you turn your own data into a revenue stream while safeguarding it like intellectual property. Please note that this information is for informational purposes only and should not be considered legal or financial advice. Consult a professional for guidance specific to your situation. Now that the legal disclaimer is out of the way, let's dive in.

1. Understand Your Data Value

Identify the types of data you generate that holds value, such as browsing habits, purchase history, health data, and social media activity. Platforms like Digi.me and DataWallet can help you understand and leverage that value.

2. Use Privacy Tools and Platforms

Privacy-focused tools and platforms enable you to control and monetize your data securely. For instance, Brave Browser rewards you for your browsing data while maintaining your privacy. Similarly, Ocean Protocol allows you to monetize your data through a decentralized marketplace.

3. Data Encryption

Ensure your data is protected by using encryption. Tools like VeraCrypt and BitLocker provide robust encryption to secure your data both in transit and at rest, preventing unauthorized access.

4. Personal Data Marketplaces

Join personal data marketplaces where you can sell your data directly to buyers. Platforms like Datum and Datumize allow you to control who buys your data and how it is used, ensuring transparency and fairness.

5. Consent Management

Consent management solutions, such as OneTrust and TrustArc, help you specify and control the permissions you grant for data usage, allowing you to retain rights and control over your data.

6. Smart Contracts

Implement smart contracts to manage data transactions. Platforms like Ethereum and Chainlink offer smart contract solutions that enforce terms and conditions for data use, ensuring compliance and payment.

7. Data Wallets

Adopt data wallets to manage, store, and monetize your data securely. Opiria and Dataeum provide this technology, which helps you maintain ownership and control over your data.

8. Legal Protections

Familiarize yourself with data protection laws and IP rights. Regulations like the General Data Protection Regulation (GDPR) and the California Consumer Privacy Act (CCPA) offer frameworks for data protection and monetization rights, helping you navigate legal landscapes.

9. License Your Data

Consider licensing your data for specific purposes. This approach is similar to how IP is licensed, granting temporary usage rights while you retain ownership. Platforms like Utopia Analytics facilitate data licensing arrangements.

10. Engage in Data Cooperatives

Join data cooperatives where individuals pool their data to collectively bargain with companies. Platforms like MIDATA and Data Union enable this collective approach, increasing the value of your data and providing better control over its use.

By understanding the value of your data and utilizing tools and strategies to protect it, you can effectively generate and monetize your personal data (a necessary first step if you want to innovate a business that helps others do so, as well). This not only maximizes your data's value but also ensures your privacy and control, turning your data into a protected, monetizable asset.

Develop a Data-driven Circular Economic Model in your Own Business

When aspiring humanity-centric innovators attempt to launch a new business with a circular economic model that uses data or some other non-traditional currency, it is advisable for them to first understand how consumer-centric innovators would approach their business, then make a few paradigm-shifting changes. First, it's important not to get sucked into the trap of greenwashing. Many companies advertise that they give a percentage of sales to charity or provide in some way for underserved communities, but simply fulfilling the bare-bones requirements of "outreach" is not humanity-centric innovation, nor does it constitute a circular economic model.

Innovators interested in developing circular-economy products or services should not start by asking the typical questions such as, "What can I sell and how much profit can I make?" Instead, the first question for any startup ought to be: "Whom can I create value for, and in what way?" This perspective can lead to more sustainable and impactful solutions that align business success with positive societal outcomes. If you already have a product or service in mind, brainstorm on how it will improve human well-being while also addressing broader social and/or environmental challenges. Overall, the idea is to align profit motives with social responsibility from the get-go.

But having the best of intentions isn't enough. For instance, to truly innovate within one of the data economies, it's a good idea to develop a product that can be used to automatically gather data and then create or align with one or more other businesses that use such data. To ensure the circular system works, you'll want to know the scope of your audience, asking how many people, businesses, or entities your initial product will touch. Finally, you'll ask, "At what stage in this cycle can value be captured sustainably?" It's not that value capture is the least consideration—far from it. You'll need profit to keep the business viable enough to do your sustainable work. But when profit is the first consideration, other elements of a circular economy tend to go by the wayside. This kind of cooperative profit-making and use of multiple currencies is, quite literally, science fiction becoming science fact. You've now entered the economy of the future.

To fully invest in the circular-economy concept, innovators should get used to pondering the following three questions before launching any product. Once asking and answering these becomes a standard part of your business protocol, you will truly be on track for fulfilling a sustainability vision.

The three questions are:

- What valuable byproduct data or materials—does the user create, and who else can benefit from them?

- Can the product be free for users while recyclers or partners pay for the valuable data or materials it generates?

- What happens at the end of its life? Can disposal or recycling not only be sustainable but also generate revenue in new ways?

8

INNOVATING A CIRCULAR, DATA-DRIVEN ECONOMY AROUND WASTE

"Owning our story can be hard, but not nearly as
difficult as spending our lives running from it."

—Brené Brown, researcher, author, and speaker on vulnerability and courage

As I've mentioned, I spent much of my chemical engineering career in the personal care business, designing toilet paper, diapers, women's sanitary products, and the superabsorbent materials that make them work. As such, the companies I worked for always did a great deal of outreach to developing nations to gain consumer insights about their needs for superabsorbent materials while also giving back through outreach and collaboration. So, as part of my job, I traveled all over the world to learn how sanitation products (or the lack of them) affect societies of all kinds. In so doing, I realized that fast-moving technology has caused many societies to develop past the infrastructure that supports them, leaving our planet in the midst of a global sanitation crisis. So, with my personal interest in circular economies, it was inevitable that I would start obsessing on a next-level idea: *What if a used product could be worth more than a new one? What if we could develop absorbent personal care products that would be worth more than new ones and traded in a circular economic model?* I became so interested in the topic that I joined an international business accelerator called the Toilet Board Coalition, whose vision included building a thriving, circular, and "smart" sanitation economy by 2030 by supporting humanity-centric entrepreneurs.[122]

I admit my wife and I had a bit of a laugh over this new venture, but what really got us thinking was how I'd introduce myself without raising eyebrows about being on "The Toilet Board." My interest in waste data recycling overcame any hesitation I might have had about embracing sanitation and women's equity as issues core to my life's work. I believe that once waste data recycling becomes a going concern, the harvested data from people of all classes and races will be of equal value. Economically, I truly believe waste data recycling could become a great economic equalizer with the potential to reduce or eliminate wealth disparity around the globe.

In the version of the circular economic model where the system recycles used products for profit, the manufacturer provides a recyclable product for free to consumers along with a system for recycling the product such as a bin and collection service. The used products are then taken to a recycling facility that extracts data from them, often for use in medical studies and pharmaceutical development. In this case, the recycling entity pays the manufacturer for the used products, and the users themselves pay "nothing," although they're actually paying with data they themselves don't consider valuable and will not miss.

There are not yet a lot of examples of profitable, waste-recycling circular economies in action, but of all the revolutionary new economies possible, this one is closest to my heart. I think it can address the two UN SDGs to which I've dedicated my professional life:

> #5: Gender equality: achieve gender equality and empower all women and girls.

> #6: Clean water and sanitation: ensure availability and sustainable management of water and sanitation for all.

The point of such paradigm-shifting, circular business models is to return economic power back to the general population. A humanity-centric business is not built to exploit people of any economic class but to uplift those at the bottom of the pyramid (at nobody's expense) and make long-term environmental gains. Still, the business must profit whether or not it saves the world, or it won't be saving the world for long, thus potential entrepreneurs in this field must brainstorm on how their products can be used to produce another currency, like data, that some specific industry will find valuable.

In some instances, people pay more for products that advertise themselves as being sustainable, biodegradable, or recyclable, because such items appeal to wise "green" consumers. While it's positive to tap into humanity's goodwill, making "doing good" primarily accessible to wealthier consumers limits its impact and inclusivity. Studies show that the average consumer simply is not willing to pay more to do good.[123] So, for a business to truly thrive, its sustainable products should cost less (in traditional currency), not more, which requires finding ways to profit from recycling. The object of a circular economy is not to supply the world with more-expensive, eco-friendly products but to bring sustainability to the masses by making green products the preferable, more economically viable choice. This should not be as difficult as it sometimes appears to be, because if you take the long view, sustainability is free. When we design products, solutions, and systems that use less energy, less water, fewer materials, and generate minimal waste, costs naturally decrease over time. Famed author and businessman Philip Crosby championed this idea way back in 1979 with his book *Quality is Free*, where he made the point that when you "make it right the first time," the savings naturally follow. This principle still holds true today.

The Circular Economy that Monetizes Agricultural Waste

One established method for profiting from waste is found in agriculture, where many farmers innovate by finding markets for their non-edible waste products. The classic example is horse manure, which can be reused or resold as fertilizer or garden compost. Another innovation was found by buckwheat growers, who used to simply dispose of the buckwheat hulls sloughed off in the manufacturing process. But these hulls were eventually found to be excellent stuffing for firm pillows such as meditation pillows, and now buckwheat hulls are sold at a premium to boutique purveyors of meditation supplies. The hulls also make excellent mulch, so farmers now bag up this "waste" and sell it to garden supply stores all over the world.

In the same manner, cellulose-rich corn stalks, once considered waste, are now sold for use in making ethanol, a gasoline substitute. Unproductive oil seeds, such as soybean, canola, sunflower, and jatropha, can be recycled into biodiesel, which is increasingly in demand as an alternative fuel. After sugar is extracted from sugarcane, the leftover stalks, known as sugarcane bagasse, can be used to create a biodegradable plastic[124] that makes a suitable substitute for conventional plastics like expanded polystyrene (Styrofoam®), polyethylene (PE), polypropylene (PP), and, in some cases, polyethylene terephthalate (PET), making them a sustainable alternative for items like food containers, cutlery, bags, and packaging.

Having let farmers take the lead in turning dross into gold, the easiest way for other industries to build their own versions of circular economies is to brainstorm on ways to reuse their own waste, especially through inter-corporate cooperation, such as the "closed loop" system mentioned earlier, where the heat and steam emitted by industrial plants are used to power other industries.

Ideas for a Circular Economy Around Human Waste

My background in innovating superabsorbents has focused on its use in personal care products, which have gone through so many advancements in the past couple of decades that I'd call it a revolution. Everyone in the developed world has access to such products. As a matter of course, families stock their bathrooms with soaps and cremes and tooth-care products and first-aid kits and tissues and hygiene items of all kinds. Most folks take such things for granted, but I know how much engineering goes into super-soft and absorbent toilet paper and highly efficient wound-healing products, and it makes me aware that there is so much more innovating that could be done.

We live in a world where every computing device is a multi-functional machine. Our computing lives are hyper-organized and efficient; meanwhile, in middle-class American homes, bathroom drawers and cabinets are often overflowing with a vast variety of personal care products, many of which per-form redundant tasks. This dichotomy seems out of step with the times and has made me wonder why we can't streamline these personal care products into multi-functional items that extract biodata and biomaterials to give us information about our health. Doing so would enable us to recycle our waste for the sake of medical technology to make the world safer, while transforming our bathrooms from cluttered spaces of excess into hubs of intelligent, purpose-driven design.

This type of innovation would also create new revenue streams both for the companies and end users of the products while potentially providing for the needs of underserved populations that otherwise would lack access to basic medical tests. Users who recycle their used products could earn rewards, monetary incentives, or direct revenue from their harvested biodata or biomaterials.

Such ideas might sound like something out of a sci-fi movie, but I've been in engineering innovative products all my life, and I feel confident this kind of thing is possible for the right company—one that sees a way to profit from fulfilling the international need for sustainability, health, waste-reduction, and economic inclusion. The following circular-economy ideas might sound a bit strange, but biodata valuable to medical science can be extracted from all kinds of things that typically go down the toilet. This type of innovation is not as advanced, yet, as it should be, mostly because of social taboos and the "ick factor." People don't like to talk about human waste. But a great deal of our waste contains valuable biological materials that can save lives—our own and those of others. I've stepped away from personal-care-product innovation, but if these ideas spark something in you, consider this your challenge—take the ball and run with it. The future is yours to shape!

Detection and Collection of Biodata from Women and Babies

Cord-Blood Stem-Cell Banking

Stem-cell-rich umbilical-cord blood, which is often discarded after birth, holds incredible potential to save lives and improve health outcomes.[125] Mothers, particularly in underserved areas, could benefit from access to programs that recognize the value of this life-saving resource. While cord-blood banks exist, they mostly operate on a donation-only model, and so much of this precious resource goes unused. By creating a circular economy model, cord blood could be collected ethically and compassionately: a for-profit cord blood bank might offer free comprehensive OB/GYN services to expectant mothers in exchange for cord blood, which can be supplied to medical researchers, pharmaceutical companies, and hospitals. This approach respects and empowers mothers, providing access to essential healthcare while contributing to advances in life-saving medical research.

Adult Stem-Cell Banking from Smart Feminine Hygiene Products

For a long time, it was believed that medically valuable stem cells could only be extracted from human embryos or umbilical cords. However, adult stem cells also exist, and recent

studies show they can be collected with remarkable ease.[126]
More specifically, they're called endometrial mesenchymal
stem cells (eMSCs), a type of stromal cell found in the endometrium. They are considered 'multipotent,' meaning they
can be coaxed through scientific processes into becoming
fat cells, bone cells, cartilage cells, and possibly even smooth
muscle cells. Ongoing research suggests that one day, medical science may harness these cells to repair tissue, treat
diseases, and even save lives. As I've noted, with today's AI
moving science forward faster than ever before, we're already
on the cusp of this and many other world-changing medical
breakthroughs. Luckily, these cells, which grow in the human
endometrium, are naturally expelled in menstrual blood.
That said, I wonder if a company could be founded that
distributes menstrual pads to communities and also collects
them for recycling, but, instead of charging the users for the
service, it would actually distribute and collect the products
for free while ultimately receiving payment from a stem cell
bank willing to pay for the endometrial stem cells that can
be extracted during the recycling process—a perfect circular
economy. Such products could promote health, dignity, and
inclusion by ending menstruation-based hardships experienced by disadvantaged women the world over.

As far as I know, only a couple of companies, Cryocell and
Life Cell, currently offer an adult stem cell banking service,
but these do not serve needy communities, necessarily. They
are simply facilities where individuals pay to store their own
adult stem cells for potential use in medical cures that do not
yet exist. Private endometrial stem cell banking like this is a

start, for sure, but, to my mind, it's a backwards economic model. Seems to me, because endometrial stem cells are so potentially valuable, abundant, and easy to gather, women producing the resource should receive something in exchange for making them available to the larger scientific community. The idea of women monetizing their own endometrial stem cells promises yet another new form of currency made possible by the rapid progress of medical science.

At the same time, I think it's possible these recyclable menstrual pads could be made into "smart" products able to monitor menstrual health by providing indicators of infection or abnormalities such as Polycystic Ovary Syndrome through color changes in the product or messages sent to an app. There is really no end to how technology can contribute to gender equity. All we need are clever, tech-savvy innovators with imagination to spare and the spunk to found and fund taboo-breaking companies.

Meconium and Micro-biome Harvesting

The first bowel movement of any baby mammal is a special substance called meconium, which is composed of materials ingested during gestation. In medical research, meconium holds potential value for studying prenatal health, as it can reveal insights into environmental exposures, maternal health, and fetal development.[127] This rare sample provides a one-time glimpse into the prenatal environment, offering valuable data that can inform advances in neonatal and maternal healthcare. While primarily valued for its scientific contributions, meconium's potential for early diagnostic research and understanding developmental health is highly respected in medical science. Why not develop a diaper distribution-and-recycling program, but in this case, the recycler would be the one to pay the diaper manufacturer/ distributor for the valuable meconium found in the used product. Meanwhile, the users in developing nations would get their diapers for free. Certainly, the recycler would have to process a lot of post-meconium diapers to provide a comprehensive service to mothers, but AI can help automate that service. The recycler would have to educate mothers as to the value of that first meconium diaper and provide a special recycling bin for it. If meconium is valuable enough, there would be profit in this circular-economy idea.

Recyclable diapers could also be designed to collect other types of infant microbiome data and materials that can be used to develop treatments for conditions like irritable bowel syndrome in adults. It seems to me, there might be innumerable ways diapers could contain valuable data and biomaterials that contribute to medical research and innovation.

"Smart diapers" could also easily be used for biodata detection. They could monitor the baby's health indicators, either by changing color when certain abnormalities are detected or even by sending a signal to an app. This way, parents and healthcare providers would be kept constantly up to date, maximizing healthcare and alleviating stress for already-hardworking caregivers.

Predictive Baby-Biodata Detection

The truth is, babies only need a diaper for about fifteen minutes a day, yet they're wrapped in one around the clock. This constant exposure often leads to diaper rash and skin irritation. When a problem's solution causes a secondary problem, there is always room for more innovation. After all, the purpose of diapers isn't just to make babies *look* comfortable. To this end, a truly revolutionary idea in the world of diapers utilizes smart technology for *predictive* biodata detection to facilitate parental intuition and almost do away with diapers altogether. The idea of this app-based diaper system is based upon an ancient practice called "elimination communication,"[128] which enables parents to recognize and respond to their baby's natural cues for bathroom needs before the baby eliminates.

In ancient days, elimination communication made diaper rash altogether nonexistent,[129] but the technique, involving near-constant contact with the baby, is inconvenient for modern lifestyles. However, when combined with predictive smart apps that signal when a baby is about to need a change, this ancient and highly effective toilet-training method could be made viable again. The app would enable parents to diaper only when necessary, thus prevent rashes, reduce waste, and bring innovative next-level thinking to infant care. This small shift could have big potential for families and their lifestyles, illustrating how even the most routine practices can be reimagined for a smarter, healthier future.

Biodata Detection and Collection from Household Products
Biodata from Saliva

Saliva is becoming an increasingly preferred method for collecting health information known as biomarkers, which provide information about inflammatory conditions.[130] Blood collection is the traditional method for such collection, but, by comparison, saliva (which contains numerous important proteins and peptides) is more accurate, less invasive, has fewer safety issues, does not clot, and requires no specialized personnel to collect. Saliva can test hormone levels, drug presence, and antibodies, all of which are biomarkers for fertility cycles, infectious diseases, and drug levels. In fact, as opposed to blood tests, saliva tests are now considered more likely to facilitate early detection of disease, but this aspect of medical science is still in its infancy.

It seems to me that with the new, rapid surge of progress in medical science fostered by AI, today's engineers should be able to develop a method for collecting data from common household items such as:

- toothbrushes
- water bottles
- disposable cutlery

Such "smart" products would sense the presence of saliva and react accordingly (by, for instance, turning different colors, like litmus paper) thus providing users with immediate information as to the presence of disease, fertility cycles, or drug over- or under-dose. Such built-in home tests would eliminate the need for visits to medical clinics as well as needle punctures and the resulting biohazard waste—all social and environmental benefits. What's more, such disposable products could also be collected for recycling by biodata harvesting agencies, where the data they provide could be studied for the enrichment of medical science.

Biodata from Facial Tissues and Bandages

When scientists use nasal cytology science to view mucous secretions under a microscope, they're able to diagnose conditions such as allergies, rhinitis, and infectious diseases with a great deal of detail.[131] It seems to me, the next step in nasal cytology ought to be the development of facial tissues that automatically perform such tests to identify biomarkers that indicate whether a person has the flu, allergies, or COVID-19. Collecting these data could help in early diagnosis, reducing the burden on healthcare systems.

Using the same paradigm of using personal care items as biomarkers, bandages could function in the same way. If they were designed to automatically test blood for various biomarkers, bandages could provide a signal to indicate important health information that would indicate the need for a trip to a doctor or specialist. Bandages could also be analyzed for biodata through a recycling system, where hospitals could send batches of this type of bio-waste to a central collection agency that analyzes the biomarkers to develop insights for furthering medical science.

My experience engineering feminine hygiene products has made me realize that the next step beyond the advancement of comfort and convenience in women's products is to make them multi-functional. I envision a pantyliner designed to change colors or other signals when it detects biomarkers indicating sexually transmitted disease, pregnancy, or uterine infection. For women in the developed world, this would be a great convenience, but for women in BoP populations, where simple medical tests can be difficult to come by, it would be a life saver. In fact, it seems to me, the same technology could be used in an adult diaper or underwear that could serve as early detection for colon cancer. Integrating technology into our ordinary personal care items is a great way to take advantage of the rapid advancement of medical science wrought by modern technology. Everybody wins. Could your startup be the first to make such revolutionary products into the next advancement we all take for granted?

Existing Smart Personal-Care Products

Many smart personal care products already exist, which shows us technology is on track toward developing more of the same. Each of the following products uses sensors of some kind to inform users of useful health data. Ideally, however, as smart technology becomes more affordable and ubiquitous, these products should not be reserved for those who can afford the top-of-the-line but made affordable and available to the public and to those in developing nations, who really need this type of in-home health monitoring. But there is even more such devices could do, such as collect valuable (and trade-able) data on users' skin, hair, general health, and personal care routines.

In order to be truly humanity-centric, improving this type of technology will require the development of improved data privacy technology along with navigating current, entrenched regulatory hurdles to benefit those at the Bottom of the Pyramid. What's more, in an ideal circular economy, the following devices would also be designed for easy disassembly and recycling, with components that could be reclaimed for reuse in new products. With the following smart personal care products, innovation is already halfway there. Could you be the one to take personal care products to the next level?

Smart Facial Cleansing Brushes

The FOREO facial cleansing brush collects data on skin type, usage frequency, and cleaning patterns, all of which help users optimize their skincare routines while providing valuable feedback for product improvement.[132]

Smart Razors

Smart razors could track shaving frequency, blade usage, and skin sensitivity to help users achieve a better shave. This data could inform the company about product wear and consumer preferences.

Smart Hair Care Products

Smart hairbrushes from companies like L'Oréal's Kérastase Hair Coach measure hair quality, brushing patterns, and scalp health. The data helps users maintain healthier hair and guides product development.[133]

Wearable Health Monitors

Wearables like fitness trackers and smartwatches collect extensive health data including activity levels, sleep patterns, and vital signs. This data supports better personal health management and informs healthcare providers of essential medical data.

Implementation of Human Waste Recycling Systems

To develop smart waste-recycling products, a great deal of money and time will have to go into research and development around the concept of extracting human waste—be it saliva, blood, or mucous—from recyclable products. This concept will also require collaboration between medical and techno-logical experts as well as a robust logistics model around a convenient method for distribution, collection, processing, and recycling. Finally, to ensure a steady supply of products to the system, consumers will have to be educated about its benefits, including the incentives and rewards they'll receive for participating. A lot of engineering is required for systems this complex, but, once upon a time, cities never dreamed they'd have efficient recycling programs for bottles, cans, and plastics like they do now. In fact, the notion of central gar-bage collection itself once seemed like a pipe dream. Making such dreams into reality is the essence of humanity-centric innovation and the very idea of it changes how we live in the world. In fact, the biggest compliment to any innovator is to have society get so used to its revolutionary products and services that it takes them for granted.

9

IMAGINATION, COURAGE, AND A WILLINGNESS TO BREAK TABOOS

Imagination is more important than knowledge.
Knowledge is limited. Imagination encircles the world.

— Albert Einstein, theoretical physicist best known for developing
the theory of relativity

Naturally, the better technology gets, the more options we have for innovative ways to recycle what used to be waste products. In and of itself, the act of capturing waste like steam for re-use elsewhere requires a great deal of technological innovation. Luckily, we have technology today that is not only incredibly advanced, but, with AI, it regularly improves itself. Technologies such as genomic sequencing, blockchain encoding, augmented and virtual realities, additive manufacturing, robotics, and all the advantages of the internet itself make what used to be pipedreams into realities. In fact, the bold visionaries who dare to dream the impossible and challenge the limits of reality are the driving force behind humanity-centric innovation. It's the audacious thinkers— the ones unafraid to turn 'pipedreams' into possibilities—who ignite the breakthroughs that redefine our future.

With the technology we have now, there is really no reason not to let our imaginations really fly. Today, staying grounded in what's possible is foolish, because what's possible changes constantly. Modern innovators aren't held back by the limits of reality and existing technology, as nowadays "reality" is a moving target. In fact, the one thing technology can't do for us is dream. That's right, we still don't have robots with imagination.

Einstein himself saw the current industrial revolution coming. When he observed the forward march of technology in his day, he said something that is just as true now as it was then. Before the internet, Sir Francis Bacon was right, "knowledge is power."[134] But as technology processes and analyzes more and more information, knowledge becomes a commodity available to everyone. What's valuable today, Einstein told us, is imagination. The ability to dream of what doesn't yet exist, the ability to aspire toward the impossible and indulge in wonder—that's where true power lies.

Along with that, I'd add that one more element is necessary for humanity-centric innovation to really get across the finish line and save the earth for humanity: the courage to act. It takes courage just to be imaginative, just to put your

dreams out there unashamedly and go ahead and be impractical. It takes even more courage to risk business capital and
act. Furthermore, it takes top-level courage to go all-in on
innovation. That courage, paired with imagination, may be
the most valuable resource for any humanity-centric innovator today.

I don't mind saying my own imagination has been the
engine behind much of the innovation I've fostered. Still,
though, many of my world-changing humanity-centric ideas
remain in the dream stage. The one thing in common about
all of them is that they break social taboos and require a bold
growth mindset. Innovators who seize such ideas will be
poised not only to make a difference on this earth but also
to make a tidy profit dealing with something "icky" nobody
else even wants to talk about. The key to success, I think,
is to really relish breaking taboos so that you can be a true
disruptor and iconoclast who is far ahead of his or her time.
That, I believe, is how certain members of Generation Z will
become the household names of their generation.

A Taboo Revolution

Through my work with the Toilet Board Coalition, a
sanitation-economy business incubator, I learned a lot about
the lives of women in developing nations. Much of the information broke my heart, but it also gave me a much-needed
education about exactly how social inequity begins. Often
taboos around menstruation and other types of human waste
lead to women lacking access to necessary products. Even

where such taboos don't exist, a lack of affordable products can cause the same result: women's freedom is restricted.

Furthermore, I learned that in some countries women's public bathrooms are either nonexistent or unsafe. As a result, many women suffer from urinary tract infections due to having to hold their water too long when they're away from home. To add insult to injury, diagnosis and treatment of such infections is often difficult to get. I learned that when girls in poverty get their periods, it often destroys their access to education beyond elementary school, because without a ready supply of sanitary pads, they simply can't be out in public for part of every month.[135] For the same reason, some women, even if they're educated, often can't hold regular jobs. This situation makes girls and women (and the children they often care for) house-bound and often susceptible to domestic abuse and other problems affecting those lacking control over their lives. The same holds true for women with babies who lack diapers.

Certainly, different cultures have created their own versions of things like sanitary supplies and baby diapers for centuries, but modern society requires its citizens to be more mobile and independent than ever, and the "old ways" of dealing with sanitation are based upon old-fashioned life-styles, and they don't work well with many of today's careers and school systems. What's more, some women's inability to participate in modern society for a lack of sanitary supplies holds back their families and communities financially, exacerbating economic disparity, which historically has been the world's greatest cause of war. Universal access to better sanitary products and diapers can help solve a whole raft of problems . . . but only for those who have access to these products and can afford them. This is where humanity-centric innovation with a circular business model can really save the day.

Make no mistake, whether you're a man or woman, and no matter what country you live in, these issues affect you and the world you live in. Women are more than half the world population, and their ability (or inability) to contribute financially to their families, and in many cases to be the family's sole financial support, is a major driver of world economies and a major player in economic recessions. Economic recessions cause high interest rates, reduced wages, increased national debt, stock market slumps, and job losses. These events tend to bring on rampant social inequity, which is the primary cause of the unrest that leads to war. I feel it is not overstating the fact to say that this taboo issue, known as "period poverty," lies secretly at the heart of many failed efforts toward world peace.

Future humanity-centric innovators, take note: the potential to develop circular economies around the distribution and recycling of menstrual products is wide open. This area of business innovation is not just close to my heart and a subject of my scientific interest, it's a field that's underdeveloped and full of potential for young innovators.

Period Poverty

For those in the know, COVID-19 and the resulting poverty and social isolation brought the issue of women's access to menstrual products to the forefront. Starting in 2020, new lifestyles and deprivations required many women managing families to allocate money only to items deemed "essential," and where men were making lists of essentials, menstrual products were left off. Because of taboos around discussing such things, husbands and fathers failed to comprehend the dramatic way in which this lack would limit their wives' and daughters' abilities to thrive. The same taboo kept many women quiet on the issue, further exacerbating the private pain experienced that year.

Period poverty is defined as a lack of access to hygienic menstrual products during monthly periods, inaccessibility to basic sanitation facilities, and a lack of education regarding menstrual hygiene. It's an issue that affects countries in humanitarian crisis but also developed nations. Statistics reveal that 10 percent of girls in the UK have found themselves unable to purchase menstrual products.[136] Not all schools and workplaces provide clean toilets and water, and many women can't afford the menstrual supplies that ensure freedom from infection and discomfort.

The taboo is such that the 2014 United Nations Educational, Scientific, and Cultural Organization determined one in every ten menstruating adolescents misses school during their menses due to insufficient accessibility to menstrual resources.[137] It is also common to develop stress, social isolation, depression, and anxiety due to the stigma around menstruation itself as well as inadequate privacy, unclean washing facilities, and a lack of effective products for managing this natural process. Furthermore, when girls experience abnormalities in their menstrual health, they often lack access to relevant medical care, and long-term health problems can result. Yet, there may be nobody to turn to for solutions.

If you recall chapter four, *What Innovation Looks Like, Today*, you'll remember that innovation begins with identifying a consumer pain point. Well, if worldwide period poverty isn't a consumer pain point, I don't know what is. Of course, the problem manifests differently in different cultures, within different socio-economic classes, and within each unique generation of women. Interestingly, from a business perspective, these differences are a bonus, as they provide opportunities for start-ups to develop uniquely customized products for different market niches.

I have worked for major US corporations and traveled the world for them, studying the issue of period poverty in depth to develop products appropriate to solving the problem in each market, from G7 nations all the way to the developing world. And yet, I was unsuccessful in my goal of developing a system of recyclable menstrual supplies, a method by which to harvest endometrial stem cells from used menstrual products, or a way to sell used menstrual products for medical research. Thus, these innovations are still on the table, waiting for a clever, unabashed, and ambitious startup to seize the day.

Helping women and girls in the developing world escape period poverty necessitates building a circular economy. To reiterate an idea mentioned above: such a system would offer free products to women and girls, receive payment in the form of something they don't value (like used menstrual products), sell that to someone who sees value there (like a pharmaceutical company, stem cell bank, or medical research facility), and receive payment from the recipient for managing the circular system. Once forward-thinking entrepreneurs get used to thinking about business in this circular way, and once brilliant young innovators realize humanity-centric work is the only truly fulfilling life choice, all the pieces of the puzzle of modern innovation will fall into place. Other opportunities that address the same and similar pain points lie in menstrual education, public toilets, baby changing stations, and providing clean water to remote and polluted areas.

The World Economic Forum estimates that it's going to be almost three hundred years before we close the gender gap economically, a hundred years before we change it politically, and a little bit more than a decade to change it from an educational perspective.[138] And all these numbers have gone up since COVID-19. I find it amazing that the US space program managed to send a rocket to the moon in less than a decade, but we, as a country, can't seem to apply that same sense of urgency to solving gender and race-based inequality and injustice. It is my plea for you, the new generation of changemakers, to make this your cause. The path I suggest (although it is by no means the only path) is the application of a circular economic model to business ideas that address humanity-centric pain points and defy social taboos. This technique, I believe, is also the key to your innovative entrepreneurial success.

Nature Has No Taboos

No matter where you live, every society has topics that people tend to avoid. Some taboos are nearly universal, while others are unique to specific cultures. For example, in Germany, discussing personal religious beliefs in casual conversation is uncommon. In China, conversations about death or the number four—associated with bad luck—are often side-stepped. In many Western cultures, asking someone's age, especially a woman's, is considered impolite, whereas in Vietnam, it's a common and expected part of introductions. Social norms around body image also vary; in some cultures, a fuller figure is seen as a sign of prosperity and good health, while in others, discussions around weight can be more sensitive.

We often follow these unspoken rules to keep interactions smooth, but sometimes, avoiding certain topics can limit progress and innovation. A great example is the hesitation to discuss human waste—a subject that's essential to public health and sustainability but remains taboo in many parts of the world. Having been in the business of superabsorbents and personal care products for most of my career, I've become very familiar with all aspects of collecting human waste from females, males, the old, and the young, and I talk about my fascinating work all the time. There are, indeed, times when my wife must nudge me in public and give me a look that means, "Babe, not everyone wants a TED Talk on bodily functions right now." My immersion in this work has, for me, eliminated this taboo. Frankly, I think it's a good thing because dealing with human waste is still a big issue all around the world, and as changing technology changes societies, infrastructure changes, as does the availability of various essential manufactured products. When it comes to waste, this is a much bigger issue, internationally, for women than men.

As lifestyles for women change, their access to affordable sanitary products must change as well, or they suffer a loss of health, safety, and opportunity. Women working in the corporate world can't use the same types of sanitary products as women living in pioneer-era log cabins did. It just wouldn't work. Similarly, those caring for infants today wouldn't be able to live a modern, on-the-go lifestyle using the leaky, ineffective diapers of the olden days. Sanitation products are a very important part of our lives. In fact, they're a big part of what provided our society with the freedoms it enjoys, today. Yet, we don't like to talk about such products, and for that reason we're missing out on incredible medical breakthroughs as well as opportunities for eliminating landfill waste during an era when both are sorely needed.

But in the natural world, where nothing is intrinsically good or bad, there are no taboos. Whether underwater, in a forest, or on a sandy plain, everything that grows, dies, or occurs is conducive to some form of life. Animal feces become compost for plants, plant matter becomes food for animals and insects, and the death of one entity produces essential chemicals required for the growth of others. Even carbon, which we tend to see as an enemy of a healthy atmosphere, is not, at its core, bad. Carbon, after all, is the building block of life. Carbon is part of carbon dioxide, which enables photosynthesis. It isn't bad to have carbon in the Earth's atmosphere until it gets out of balance with the other atmospheric elements.

Rather than categorize any of its elements as good or bad, nature merely attempts to seek balance between them, to ensure the cycle of life continues for all species. In fact, if you think about it, if the excess carbon in the atmosphere causes humanity to go extinct, we will stop producing carbon, so, even if it takes millions of years, the atmosphere will be able to achieve balance again. That's nature, in its infinite patience, working to seek planetary balance. But if we humans can help achieve that balance in a different way, our species will be both cooperating with nature's goals and assuring our own survival—the ultimate combination of natural processes at work.

Humans, however, have left the natural order in the sense that we tend to view certain aspects of the world as "good" or "bad." We've created taboos to keep ourselves safe (and away from icky things), yet nature doesn't need taboos to keep its animals healthy and societies functioning. So, I think it's safe to assume we humans must have taken a wrong turn somewhere. Take geysers, for instance. These muddy, violent, hot, rotten-egg-smelling natural phenomena perform the essential function of releasing geothermal energy from the earth's core. Along with that heat comes sulfuric acid and hydrogen sulfide gas. A geyser erupting is basically a planetary fart. It's gross, but it's not bad. It's necessary. Mangrove swamps emit a similar disgusting smell, but they also sequester carbon from the atmosphere at a rate ten times greater than mature tropical forests. We need those disgusting things. The more the better! The female praying mantis bites her lover's head off after copulation, but she needs that extra protein to lay more and healthier eggs. Some clever males, however, escape the ritual decapitation. Since they're the smarter and stronger males, they'll mate more often and pass on more of their superior genes. I'm glad I'm not a praying mantis, but none of these natural phenomena are intrinsically bad. Nature doesn't classify things that way. It just functions for the survival of every species with whatever tools are at its disposal. In the same way, we humans would benefit from ceasing to classify the most unpleasant elements of our world as bad and therefore taboo and therefore unlikely foci for innovation.

Biomimicry

Biomimicry is a style of systems thinking that steps outside our culturally created norms and simply seeks to learn from nature by valuing everything that exists as an essential aspect of natural growth. Whether we speak of forests, savannahs, oceans, deserts, or Arctic ice floes, nature always integrates waste, death, and reproductive processes into its system without judgement or bias. Everything that exists, even if just for a second, is an integral part of a huge, never-ending, waste-free cycle. Left to its own devices, nature has no trash dumps. Rather, every inch of the natural world is a built-in recycling center. Biomimicry is the science of examining the genius of the natural world and finding ways to innovatively imitate its principles in our industrialized world.[139]

A common example of biomimicry is the ubiquitous Velcro strip. In nature, burr seeds attach to mammal fur in a hook-and-loop configuration. This method of hitchhiking enables the plant to spread its seed far and wide, ensuring the survival of its species. Swiss engineer George Denistral, Velcro's eventual inventor, noticed burrs attaching to his clothes on a hike, and when he put one under a microscope, he discovered the unique hook-and-loop system.[140] Then, he spent years attempting to biomimic this attachment technique until he invented Velcro: "the zipperless zipper." Velcro got its big break in the 1960s, when the fastener was utilized by NASA for elements of space suits. Then, in the 1980s, talk show host David Letterman became so intrigued by the fastener that he constructed a suit and wall of opposing sides of Velcro and wore the suit in order to jump onto the wall, where he stuck in mid-air quite dramatically. Now, Velcro is an effective fastener found in millions of items all over the world—all because someone looked at a common natural "pest" as a phenomenon instead of just an inconvenience.

The essentials of biomimicry were first identified by Janine Benyus in her 1993 book, *Biomimicry: Innovation Inspired by Nature*. These principles have been extensively studied and were eventually codified into twenty-six principles used by all organisms and ecosystems on planet Earth to create conditions conducive to life. These principles are not only excellent guides for all types of engineering and design, but philosophically they're central to effective systems thinking, itself. You can think of the principles of biomimicry as super-tools for innovation.

Years of research have boiled the essentials of biomimicry down to seven essential life principles, each with its own sub-topics, for a total of twenty-six principles:

The Principles of Biomimicry

- Evolve to survive
- Replicate strategies that work
- Integrate the unexpected
- Reshuffle information
- Adapt to changing conditions
- Incorporate diversity
- Maintain integrity through self-renewal
- Embody resilience
- Be locally attuned and responsive
- Leverage cyclical processes
- Use life-friendly chemistry
- Use feedback loops
- Cultivate cooperative relationships
- Be resource efficient
- Self organize
- Build from the bottom up
- Combine modular and nested components
- Integrate development with growth
- Use low energy processes

- Use multi-functional design
- Recycle all materials
- Fit form to function
- Use life-friendly chemistry
- Break down products into benign constituents
- Build selectively with small subset of elements
- Do chemistry in water

The above list is enough food for thought to inspire any engineer, designer, or businessperson for a lifetime. It pays to look at whatever school, business, or even artistic endeavor you participate in and ask yourself if you're using the seven primary principles in your work. If not, the keys to improving may lie above, in the principles of nature itself.

For inspiration, consider Dr. Frank Fish, a biology professor and biomechanics expert who was surprised to notice bumps (known as tubercles) on the leading edge of a humpback whale's fin.[141] Like most people, he assumed effective fluid dynamics required smooth surfaces to cut through the water, so he couldn't imagine why a whale would have these bumps. Curious about the phenomenon, he joined forces with a mechanical engineer to further investigate. Indeed, they found that tubercled blade designs reduce drag and improve lift—one of the primary goals of fluid dynamics. Because the same system works aerodynamically, the pair partnered with a businessman to start Whale Power, a company that creates highly efficient wind turbines and fans.

Innovate with Biomimicry Using Form, Function, Process, and Context

One way to follow in the footsteps of successful biomimicry pioneers like those above is to begin, like Dr. Frank Fish, by investigating and emulating a natural *form*. But don't limit yourself; think on nano, micro, meso, and macro levels. It also helps to think in terms of the *process* of which the system is a part—after all, everything in nature is created through a natural process, functions as part of a purposeful process, and its death and decay follows a specific process as well. Finally, all forms work within *systems* such as neurological systems, ecosystems, and nutrient cycles. So, you'll learn more about the natural form that fascinates you when you view it in the context of both its process and system.

Having encouraged you to become fascinated with natural forms, I must contradict myself a little here, as it's important to note that innovating with biomimicry begins by resisting the temptation to focus purely on form. What's more important is the *function* any natural element serves. In innovation, this can work two ways: One can identify a fascinating function (such as bumps on whale fins making the whale swim faster) and look for where in your design project such a phenomenon could be utilized. Or one could take the opposite approach and identify functional problems you need to solve, then go looking to nature for inspiration. The trouble with that second technique has been that biology textbooks tend to be arranged by genus and species rather than function. Luckily, a company called Biomimicry 3.8

has solved this problem for up-and-coming biomimetic innovators.

A search engine at asknature.org enables users to look up functions such as "improving fluid dynamics" (in the case of the whale fin) or "attaching" (in the case of the burr that inspired Velcro) or "recycling waste" (in the case of my as-yet-to-be-innovated ideas) in order to learn how animals, insects, plants, and microbes have already solved those problems. This function-first biological reference site is designed for innovators working in both directions—those seeking fascinating functions to imitate and those trying to solve an existing functional problem by viewing it through nature's eyes.

Biomimicry 3.8 has further enabled your search for functions both in human life and in nature with its Biomimicry Taxonomy diagram, which divides all natural functions into groups, subgroups, and sub-subgroups, helping innovators narrow down the exact functional elements they seek to find how nature has already achieved these solutions. To experiment with the diagram on their web page, think of any problem you'd like to solve, then use the Biomimicry Taxonomy[142] to identify the exact function required of an innovation that would solve that problem.

In order to invent, design, and engineer something new, inspirational ideas must be funneled down to isolate only what is useful for your particular endeavor, and that's where *context* comes in. Having identified your biomimetic innovation's form, process, and system, you'll be better able to understand its function, but from there, you'll want to go on to define the *context* in which that function works.

In nature, the context might be large scale, microscopic scale, in the dark, at high altitude, in outer space, underwater, and so on. You'll want to take your inspiration from a natural example whose context coincides with your own. Matching *form* (including process and system) with both *function* and *context* will significantly narrow down your search for nature-inspired options. With these guidelines in place, an innovator can then find an example from nature, abstract the design principle, build the design, and continue improving it through a process of evaluation and experimentation.

Aristotle said, "Nature abhors a vacuum,"[143] referring to the idea that empty spaces are unnatural and will always be filled. In a broader sense, this reflects how none of us exists in isolation. We are all interconnected—our innovations, businesses, families, and cultures all influence and shape one another. Just as young children go through a process of individuating from their parents, it's natural for young adults to go out in search of that which they feel will distinguish them and give them meaning as separate selves, yet nature's laws prove that a separate self is an unstable entity, doomed to collapse. To want to become an innovator, you must realize how important you and your life's work are to the planet at large—and how important that planet is to your own species' survival. That said, biomimicry is an excellent way to find your own niche in a world filled with the potential for innovation. What's more, those innovators willing to imitate nature by breaking taboos have quite a creative advantage over the less-brave souls mired in conventional problem solving.

Help is Out There for Humanity-centric Innovators

Human waste recycling requires biological innovation, which sits squarely in the province of biotech and medtech. Whether you're thinking of shooting straight from an engineering college to a corporate job, as I did, graduating college with an entrepreneurial mindset, or launching your own innovative start-up straight out of high school, you'll need a good bit of infrastructure to develop the science behind your groundbreaking ideas. Many such start-ups begin at university labs, but there are a lot of reasons, after graduation, to move on from academic resources. First of all, upon graduation, you may no longer have access to the facility, and even if you don't, it's possible the university could take an interest in your discoveries or set restrictions on certain types of work, especially if they overlap with the university's affiliations or the interests of its corporate sponsors. Luckily, even if you're going it alone, you're not without resources. Business incubators and accelerators exist to provide affordable office and lab space to ambitious startups like yours. As you're setting up your business or research endeavor, it's important to know the differences between these two resources, to make the best of what's out there.

Incubators and Accelerators

In urban centers, the phenomenon of working from rented desks in a shared-office-space environment has become reasonably common for solopreneurs, gig workers, and other small-business-minded people that need little more than fast internet connections and basic office infrastructure to get their work off the ground. These facilities are base-level business incubators, but, at a higher level, some incubators provide a range of support services such as business coaching, secretarial services, start-up capital, and valuable shared equipment. For instance, incubators specifically serving biotech and medtech companies, sometimes called "life science incubators," offer wet lab facilities, which is often otherwise so expensive it serves as a barrier to entry for many brilliant young entrepreneurs. In some wet labs, the sophisticated, shared lab equipment can include state-of-the-art 3D printers, laser cutters, CNC machines, and sophisticated prototyping equipment. In that case, the incubator fee more than pays for itself in reduced investment capital. Ideally, mentorship accompanies these resources.

Business mentors can be an invaluable help to young entrepreneurs who often excel in their scientific areas of interest but may lack skills in business presentation and pitching along with knowledge of regulatory compliance. Best yet, mentors, who tend to do this work from the heart, are typically eager to provide networking opportunities for young entrepreneurs to make strategic partner connections, meet angel investors, and pursue venture capital. In fact, some business incubators are so focused on mentorship that they don't offer physical space at all, but exist in the virtual realm, providing only mentors, classes, networking opportunities, and specialized online research facilities.

That said, it's important to know the sponsor or owner of any wet lab incubator to determine if your work will be cooperative or competitive with its goals. Many such incubators are spearheaded by companies interested in furthering very specific types of science, which can end up being a big advantage, but only if you fit into that niche. Other incubators are more philosophically based, existing only to assist sustainable business ideas, community development technology, or science that furthers social equity. Then there are those who only wish to incubate ideas that further oil-and-gas-industry or pharmaceutical-industry goals. It's important to know in whose sandbox you're playing!

While incubators tend to be available to any innovator who can afford the fee, accelerators tend to be more exclusive and pricier. Theirs is a high-risk, high-reward concept. They're also more ambitious and organized, like a custom-made business boot camp for your particular innovation. Accelerators aspire to accelerate the speed of innovation by putting scientific endeavors on track to compete with the pace of industry. As such, accelerators also typically offer mentorship, office and lab space, access to unique facilities, and networking opportunities, but as part of a step-by-step program. Virtual accelerators, offering online resources only, also exist.

In business incubators, innovators are given access to facilities and can use as much or little of it as they like, on their own schedule, whereas participants in accelerators—often former academic researchers that need a crash course in turning science into profit—enroll in programs with certain requirements designed to push their businesses forward at top speed for a period of time. When the accelerator is over, participants graduate and are expected to take what they've learned and run with it, which is why business incubators can be useful for fledgling businesses both before and after participating in accelerator programs. That said, there is no law saying accelerators and incubators can't blend ideas and methodologies, so when you research these initiatives, you may find the line between the two is blurred in favor of helping innovators the best way possible.

For lists of businesses incubators and accelerators dedicated to sustainability-oriented initiatives, see Appendix B of this book.

Toilet Board Coalition Innovations

My participation in The Toilet Board Coalition, being a business incubator specifically for start-ups in the sanitation space, gave me the opportunity to function as an advisor to some forward-thinking innovators determined to solve the same problems that concerned me. Among them:

- *Be Girl*, in Mozambique, whose business branch sells reusable sanitary products to the public and whose philanthropic arm provides the products free of charge to women in developing nations along with menstrual education that helps women embrace their reproductive life and reject shame. In a true circular business model, the company would profit from some entity other than the product's end user, but that is not the case here. Instead, *Be Girl* has chosen to structure itself with equal profitable and philanthropic arms to pursue humanity-centric innovation on its own terms.

• *Padcare*, in India, which was founded to ensure women in developing nations had access to disposable menstrual pads as well as a way to recycle them. The company, whose motto is to "enable sustainable menstrual hygiene management with corporate responsibility," was founded because while Indian women have access to menstrual pads, used ones were not being properly disposed of due to a lack of waste facilities and social shame about this type of waste. *Padcare* provides a service to schools, apartment buildings, offices, and other institutions. It discretely distributes new pads and collects used ones for recycling with its recognizable *Padcare* Bins. In this case, while the items in question are recyclable, the *Padcare* service is paid for by the institution it serves, so this business also fails to meet the true criterion of a "circular" business model, but it's a step in the right direction.

While my participation as a mentor for the Toilet Board Coalition enhanced the social-responsibility initiative and brand image of the company I worked for, it also provided invaluable insights into the developing-world market, so I did good for the world while helping my company stay on the cutting edge. It's this kind of outreach that fosters humanity-centric innovation in both large US corporations and smaller start-ups, so it's a great example of how profitable business, rather than charity, is the best driver for humanity-centric solutions. Most importantly, business incubators like the Toilet Board Coalition work to promote businesses that want to employ the new, circular economic model.

10

FUTURISTIC INNOVATION
HAPPENING TODAY

It's no use going back to yesterday,
because I was a different person then.

— Lewis Carroll, *Alice's Adventures in Wonderland*

Imagine a battery that lasts for ten thousand years and could power an iPhone or electric car that never has to be charged. Imagine a world where plastics decompose like blades of grass, and petroleum is no longer relevant because artificial photosynthesis powers everything with carbohydrates instead of hydrocarbons. Imagine every medication being custom-made for its user's personal genetic traits. Imagine harvesting your stem cells in your 20s to be used decades later to cure the ills of old age. Imagine living a vibrant, healthy life . . . forever.

Members of Generation Z—and every other generation willing to listen and able to make a difference—I'm talking to you when I say: It's quite likely that, within your lifetime, you won't have to imagine these things because they won't just exist, they'll be mainstream. In your lifetime, science fiction will become science fact. Let's look at some of today's companies that are either breaking taboos or just getting wildly creative with innovation under a humanity-centric umbrella.

Nuclear Diamond Batteries

Nuclear diamond batteries (NDB) are yet another ground-breaking technology currently under development that could not only transform the world's energy grid but solve what is arguably the world's biggest waste problem.[144] Nuclear diamond batteries have extremely long lifespans, potentially lasting thousands of years, and are made using synthetic diamonds formed from radioactive carbon-14 extracted from nuclear reactor graphite. Governments and industries worldwide spend billions of dollars annually on managing and storing nuclear waste, ensuring its containment for thousands of years until it becomes less hazardous. In the United States alone, the Department of Energy allocates approximately $6 billion per year to nuclear waste management, while global decommissioning and long-term storage costs are estimated to exceed hundreds of billions over time. The idea behind NDBs is to recycle this hazardous waste product by extracting radio isotopes from it and harnessing their clean energy with a battery designed to prevent radiation leaks. A diamond battery has the potential to be produced in a range of sizes, from tiny particles to compact cells for niche applications, and could power devices continuously for thousands of years without recharging. Harnessing energy from radioactive decay within a safe diamond casing, it generates a steady, low-power output ideal for long-term uses. With its durable, self-contained design, a diamond battery could one day offer a virtually maintenance-free power source. What's more, when fully expended, its encapsulation minimizes any environmental impact. This system is known as decay-voltaic (as opposed to solar energy, which is photo-voltaic). Best yet, any excess energy the battery produces can be sold to a local power grid. Eventually, NDBs could be made in any size and won't need to be replaced, they have the potential to be in things like pacemakers and hearing aids as well as rockets, space stations, and disaster relief in remote areas of the planet. A collection of such batteries can even be scaled up to act as a mobile power plant. This technology is not currently available to the public, but it's being funded, making progress, and is sure to be a major part of the next industrial revolution.

Artificial Leaves Generate Power Just Like Plants

Researchers led by MIT professor Daniel Nocera have produced a thin sheet of semiconducting silicon they're calling an "artificial leaf."[145] It can turn the energy of sunlight directly into storeable, wireless electricity. Constructed as a silicon solar cell with various catalytic materials bonded to it, the leaf needs no external wires or control circuits to operate. Just like a real plant, when the silicon "leaf" is placed in water and exposed to sunlight, it immediately generates oxygen and hydrogen that can be collected, stored, and later fed into a fuel cell that delivers an electric current. The device, which works in ordinary water, is made of earth-abundant, inexpensive materials such as silicon, cobalt, and nickel.

Imagine using an artificial leaf to create abundant energy and pure water while helping to solve climate change. What if you could take the work of Professor Nocera's artificial leaf to separate hydrogen from sea water (or wastewater), then pipe this hydrogen to every city across the country where it would be locally combusted to produce electricity with the only by-product being pure water. An innovation like that would simultaneously accomplish three important needs for humanity:

- using the free income of the sun to produce hydrogen for energy
- desalinating the sea for clean water
- producing both energy and pure water in a carbon neutral way to mitigate climate change.

Further, imagine If you could utilize artificial photosynthesis for carbon fixation whereby carbon dioxide, water, and sunlight are used to produce complex carbohydrates for fuel, feed, food, and feedstocks for industry while removing

CO_2 from the atmosphere. After all, every botanical plant in the world does this every day, right this very moment. Why not us?

Wouldn't it be great to bring the vision of Giacomo Ciamician to life? Over a century ago, this pioneering Italian chemist envisioned a world powered by artificial photosynthesis—using sunlight to produce energy and chemicals in a way that mimics nature, without relying on fossil fuels. In 1912, he predicted a future where vast solar farms would replace coal and petroleum, harnessing the power of the sun to create sustainable fuels and materials.[146] Today, as scientists work to develop artificial photosynthesis technologies, we have the opportunity to turn Ciamician's dream into reality and revolutionize the way we generate clean energy.

Carbohydrate-Based Fuel

The need for alternatives to fossil fuels has famously spawned wind farms and solar energy arrays, both of which contribute to the world's energy demands. But after decades of development, we're still waiting for these energy sources to become scalable and consistent across all times of day, seasons, and weather patterns. Meanwhile, today's tech gurus surmise that rather than attempt to further the development of such intermittently available energy, we ask ourselves what *other* abundant, renewable, reliable, yet sustainable potential energy resources exist? After all, AI and other up-to-date technologies might completely change the game of renewable energy, but only if we open our minds to novel solutions.

As such, today's revolutionary thinkers are saying that the best way to use solar energy is not directly but secondarily by making energy from organic matter that's grown by the sun but then converts into carbohydrates and sugars. Scientifically speaking, vegetative matter, also known as biomass, is a fuel formed by using photosynthesis to convert CO_2 and H_2O into stable, high-energy, organic molecules. When we make this conversion in the human body, we're well aware that delicious carbohydrates convert easily into usable sugars that produce energy. But new data tells us some of these carbohydrates produce sugars, cellulose, and hemicellulose, all of which can be transformed into a fuel so dependable it could rival fossil fuel.

Research into this potential fuel is currently stymied by the complex chemical and/or electrochemical conversions required, but with the speed of AI data processing, these problems could be solved before long. Preliminary steps have been taken towards using carbohydrate-based fuel to power internal combustion engines, thereby making it suitable for transportation. Carbohydrate fuel cells for electrical power are also in development, where electricity would be generated by the oxidation of glucose, glycerol, or ethanol. The resulting carbohydrate-based fuels could prove to be inexpensive, non-toxic, and not easily degraded, while providing vast energy storage without time or weather constraints. Current calculations tell us that food-quality glucose can produce electricity at a cost comparable to that of today's typical electricity sources.[147] Imagine a world where fossil fuels are a thing of the past, and clean burning carbohydrate-based fuel powers everything from homes to cars to smart technology. This technology isn't a pipe dream. It's already on the way. Remember, the Stone Age didn't end because we ran out of stones, and the Petroleum Age won't end because we run out of petroleum. It will end because we will have found a better way—with your help!

Carbohydrate-Based Plastics

It was only about seventy years ago that plastics became an integral part of everyone's daily life, and now all our bodies contain micro- and nano-plastics that we simply can't get rid of. Sea creatures suffer from systemic plastic as well, while giant, floating islands of plastic garbage cause further problems for sea life throughout the world's oceans.[148] Preliminary research hasn't identified exactly how we are all affected by the ubiquity of plastic in our bodies, but we do know it's not good for humans, animals, microorganisms, or the environment. For the first time we have even found microplastics circulating in human blood.[149]

Strangely, even though we know plastic has become a huge problem, it's so convenient and inexpensive that we keep on making and using it, in spite of ourselves. In fact, we're so dependent upon plastic that if the world suddenly banned the creation of new plastic, entire societies would shut down and their economies along with them. We can no longer keep the machine of society going without plastics, but modern technology now enables us to make biodegradable plastics from non-petrochemical biomass sources such as plants, algae, animals, fungi, and microbes, which could, at least, stop escalating the damage.

I have worked on projects designed to create plastic alternatives derived from plant and microbial carbohydrates, and some versions of these plastics have reached mainstream markets. For instance:

- Biocompatible starch-based plastics—which have low toxicity, good degradation properties, and low carbon footprints—blend natural starches with thermoplastic polyesters and are used to make certain types of packaging and paper.

- Polylactic Acid (PLA) is a thermoplastic made from lactic acid derived from corn starch, tapioca, and sugarcane. At your local grocery store, you may have already bought picnic tableware made from this material, which is also useful for disposable fabrics.

I'm a big believer in the future of carbohydrate-based plastics, but this type of innovation can be complicated for a variety of reasons. One of the problems facing bioplastics innovators today is a lack of standardization and definition of the new materials, which makes it difficult to route them to the proper type of recycling or waste facility. What's more, ensuring a viable solution to the plastics problem depends upon material transparency regarding the nature of each bioplastic to ensure new products solve problems without creating new ones. So far, certain definitions have emerged that help with this, but there is still very little legal regulation to oversee the labeling of bioproducts. You may have already noticed these definitions on the packaging of various commercially available bioplastic products:

- "Biomaterial" refers to any material with a biological association. It fits nearly every material that is not petrochemical based but doesn't assure any particular level of sustainability.[150]

- "Bio-based" has been defined by the USDA and the EU to mean products derived, at least in part, from any type of biomass, but somewhat arbitrary rules determine the amount of biomass in each product that determines whether it is considered bio-based. For instance, bio-based carpet needs only 7 percent bio-content, while fabrics must have a minimum of 25 percent. For this reason, a lot of so-called bio-based materials actually have a great deal of synthetic content, and the term on packaging can be misleading.[151]

- "Biosynthetic" and "bioplastic" are also terms with loose definitions. Such items are made by converting biomass into plastics, but because literally any process can be used for that conversion, the resulting material is not necessarily biodegradable. Sugarcane, for instance, can be converted into a polymer that's chemically identical to one made from petroleum, so, while this is a bioplastic, it doesn't completely help forward the cause of sustainability.[152]

The terms "biofabricated" and "bioassembled" refer to materials grown from living cells such as bacteria, yeast, and mycelium—truly a breakthrough in sustainable innovation. However, many materials contain biofabricated or bioassembled ingredients without being entirely biological. For instance, a fabric could contain a biofabricated dye but not actually be woven from biological threads. So, again, this term can easily be misunderstood.

Further defining the terms around bioplastics will help consumers know exactly what they're buying, but to me, a more important issue is simply that innovators must prioritize sustainability as they develop bioplastics and do it for the sake of the earth and human health. With that humanity-centric ideal, the future of bioplastics is bright, but without it, bioplastics could simply become another version of the same old disposable products. That's why intent is a crucial factor of this and all humanity-centric innovations. At this point in the history of the planet, business innovation must have an idealistic component. It can no longer be "all about the bottom line."

Life-Changing Medical Innovation

If you're interested in science so cutting edge and futuristic that even *Gen Z* has barely glimpsed its potential, today's alternative medical science field will fascinate you. After all, for as long as most of us can remember, modern medicine has promised pain and symptom relief primarily through pharmaceuticals and surgery. While those techniques have advanced quite handily over the past few decades, drugs have side effects, and surgery, painful recovery periods. What's more, for those with chronic conditions, the need to spend time and money on frequent visits to far-away doctors' offices adds insult to injury. With all the science at our disposal these days, haven't you ever asked, "Isn't there something better, out there?"

Telehealth

Ever since the 2020 pandemic, people have become accustomed to seeing physicians via telehealth. Surely, you've done it, too. After all, many doctor visits are simply exchanges of information. When a physical exam isn't necessary, telehealth saves time and money for everyone involved. But telemedicine has a lot more potential than the simple act of consulting with a doctor via video call, and the power of this technology has barely been tapped. For instance, we've all known people who consult WebMD over the slightest sniffle and manage to convince themselves they have a dreaded disease. This is misuse of online medical information and it is not telemedicine, which involves personal interaction with a healthcare professional. This search is nothing more than unguided wandering in the cyber-universe of random information on symptoms. It happens because folks are plugging their symptoms into search engines. After all, while swollen lymph glands can be a symptom of Dengue Fever, there is little reason for concern unless you've recently traveled to Africa or Puerto Rico. AI can help solve the problem of rampant medical misinformation and the resulting hypochondria.

In the near future, advanced patient portals and AI-powered health assistants will seamlessly integrate with electronic health records, providing everyone—from the health-conscious to the doctor-averse—with instant, intelligent, and highly personalized medical guidance. No longer will patients have to rely on scattered internet searches or wait for office hours to get answers. Instead, they'll have 24/7 access to an intuitive digital companion that tracks symptoms, analyzes trends, and connects them to medical professionals when intervention is truly needed.

For those who regularly seek medical insights, these platforms will act as an always-available, trusted resource—one that doesn't just respond to concerns but anticipates them, using predictive analytics and real-time health monitoring. And thanks to the increasing role of genetic analysis, these AI-driven assistants will tailor advice and treatment recommendations with unprecedented precision, reducing trial-and-error in healthcare and minimizing unexpected side effects. Imagine logging into a portal that not only shows your lab results but interprets them in the context of your genetic makeup, lifestyle, and medical history—offering actionable insights before you even step into a doctor's office.

At the same time, for those who typically avoid medical care until it's absolutely necessary, these innovations will remove barriers to early intervention. AI-driven symptom checkers and virtual consultations will provide quick, accessible assessments, empowering more people to seek care sooner—often from the comfort of home. This shift won't just improve individual health outcomes; it will elevate public health by making accurate medical knowledge more widespread and actionable.

Gen Z, this is where you come in. You're the first generation to fully embrace digital transformation in all aspects of life, and now you have the chance to shape the future of healthcare. Whether you innovate in AI, data science, user experience, or bioinformatics, your contributions can make medical care smarter, more proactive, and more personalized than ever before. The next breakthrough isn't a distant dream—it's being built right now. The question is: how will *you* be part of it?

AI-Enabled Medical Devices

Diabetic children now have access to remote patient monitoring, letting parents track their blood sugar levels in real time through smartphone apps. This technology enables quick responses, preventing dangerous glucose swings. With continuous glucose monitoring (CGM) systems becoming more common, diabetes management has improved, making life safer and easier for both children and their families.[153] Similar remote sensors can also measure blood pressure, temperature, and oxygen levels. In fact, the COVID-19 pandemic would have been even worse if it hadn't been for remotely monitored pulse oximeters providing ongoing information about patient lung function.

Remote patient monitoring (RPM) enhances freedom and security for individuals with chronic conditions by enabling continuous health monitoring and early intervention, thereby reducing emergency department visits. Studies have shown that RPM programs can lead to a 25% reduction in emergency room visits and hospital readmissions.[154]

Additionally, RPM facilitates telemedicine consultations, decreasing the necessity for frequent in-person doctor visits. This approach not only conserves time and money for patients but also helps prevent conditions from escalating to acute phases, thereby reducing overall healthcare costs. For instance, the Department of Veterans Affairs' RPM program saved an average of $687 per patient per month in healthcare costs, primarily due to reduced hospitalizations and emergency department visits. [155]

While artificial intelligence doesn't have much of a bedside manner, its specialty is efficient data analysis, and nowhere is that skill more needed than in medicine

Artificial intelligence (AI) is revolutionizing medical imaging by expediting the analysis of X-rays, CT scans, and MRIs, thereby enhancing diagnostic efficiency. AI algorithms can swiftly and accurately identify subtle abnormalities, facilitating early disease detection and allowing healthcare professionals to focus on complex cases that require human expertise. This integration of AI into medical imaging not only accelerates diagnosis but also reduces the necessity for repeat imaging, optimizing patient care and resource utilization.[156]

Studies have demonstrated that AI-assisted analysis in radiology can improve the detection rates of conditions such as breast cancer. For instance, research indicates that incorporating AI into mammography screenings enhances the identification of breast cancer cases, leading to earlier and more accurate diagnoses.[157]

Moreover, AI's capability to analyze vast datasets and recognize patterns contributes to predictive analytics in medical imaging. By assessing imaging data alongside patient history, AI can assist in predicting disease progression and tailoring personalized treatment plans, thereby improving patient outcomes.[158]

While AI significantly augments the diagnostic process, it is essential to recognize that it serves as a complement to, rather than a replacement for, human medical professionals. The synergy between AI and clinicians ensures that complex diagnostic challenges are addressed with both technological precision and human judgment, ultimately enhancing the quality of healthcare delivery.[159]

AI remote sensors and image analysis have brought the medical field to a whole new level, but the potential of this technology to monitor and improve human health ranges all the way from the "I've fallen, and I can't get up" wearable panic button to real-time health monitoring for acute cardiac patients. Whatever is next in AI-enabled medical innovation, its efficient and effective use will make our planet a healthier, happier, and far more convenient place.

Regenerative Medicine/Stem Cell Therapy

Alternatives to modern, Western medicine have long existed in the field of natural medicine, herbs, acupuncture, and other options based in ancient medical science whose aim is to help the body heal itself. Such therapies run the gamut from being highly effective to lacking FDA approval and delivering mixed results. But regenerative medicine, a science that has been gaining steam for some time, combines evidence-based science with a heal-thyself philosophy. Not only does this side-effect-free medical science gracefully unite ancient and modern scientific disciplines, but it may also one day put the multi-billion-dollar pharmaceutical industry out of business.

We've all experienced getting hurt in some way and seeing how the body heals itself. Wounds scab over, cuts heal, and infections eventually go away. The body does this natural tissue repair with its own stem cells. Everyone has them. In fact, you couldn't live more than an hour without them. Typically, when a body doesn't naturally heal itself, it's due to age or overuse, where the relevant stem cells have become weak and ineffective. Regenerative stem cell therapy's answer to this is a simple injection of new stem cells that can be derived from your own body or that of another human—the younger the better. Stem cells taken from babies' umbilical cords are particularly well known for their therapeutic effects.[160][161]

Once upon a time, this therapy was politically controversial because it was thought stem cells could only be harvested from fetal tissue, but science has advanced since then, and we now have innumerable ways to harvest stem cells. Studies show we can even sometimes take them from a healthy part of our own bodies to inject them into a damaged part of the same body.

To be sure, this science is still in process. Just a few stem-cell procedures have so-far been approved by the FDA, among them: therapies for treating leukemia and lymphoma, osteoarthritis, Crohn's disease, and blood cancer. Stem cell research and therapies conducted internationally have gained popularity; however, the lack of consistent global regulations has led to the proliferation of unproven and unregulated treatments, posing potential risks to patients.[162]Still, pioneers of this science believe such therapy will soon replace expensive and painful joint replacement surgeries. Some suggest the pharmaceutical industry doesn't want this medical innovation to gain ground and put it out of business, but today's science is soon to be in the hands of Gen Z, and you'll be the scientists making that call.

The key to regenerative medicine is a little thing called the mesenchymal stem cell (MSC), which is like the conductor of an orchestra of the body's natural, self-healing ability. While some stem cells can only specifically heal the areas where they originate, the newly discovered MSC is more like a personal drug store. It assesses what type of healing is needed in any specific area of trauma and responds intelligently, naturally releasing chemicals appropriate to that situation.

Potential uses for MSC therapy include a wide range of disorders involving inflammation and immune-system breakdowns.[163] Wear-and-tear-type injuries, such as sports injuries, have also been known to respond well to MSC therapy. So, further research and development could, potentially, revolutionize the world of both amateur and professional sports with faster healing times and improved performance.

Doctors know, however, that true healing must integrate the whole self. Ideal nutrition and great stress management are part of every effective healing process, as are great lifestyle choices and living a life filled with a sense of purpose and meaning. So, humanity-centric innovation in this field not only improves medical science but could make innovators themselves healthier due to the ongoing satisfaction of such inspiring work.

Achieve Immortality

Every decade, advances in medical science have gradually increased the human lifespan to the point where now, ambitious investors are asking, "Is there a limit to this?" After all, advancements in medications, genetics, robotics, and nanotechnology reveal there may be a variety of ways to extend the human lifespan, pretty much indefinitely. Futurist Ray Kurzweil has introduced the concept of "longevity escape velocity," suggesting that by 2030, medical advancements will extend human life expectancy by more than one year for each year that passes.[164][165] This means that individuals who reach this point could potentially "live long enough to live forever," as each year brings medical progress that outpaces aging. In fact, many of today's billionaire investors have doubled down on seeking immortality, the ultimate prize. And the field is wide open. OpenAI's CEO Sam Altman has invested in Retro Biosciences, whose mission to add ten years to the healthy human lifespan hinges on the anti-aging drug Metformin. Amazon's Jeff Bezos has entered the longevity game with an investment in Altos Labs' initiative to reverse disease with cellular rejuvenation re-programming.[166] Investors Yuri and Julia Milner have also put significant capital behind both Altos Labs and the Milky Way Research Foundation, which grants funds to labs developing longevity and anti-aging therapies.

Finally, in a truly revolutionary twist, Russian billionaire Dmitri Itskov is best known for his 2045 initiative, which is building a scientific community around the field of life extension, but this organization's focus is brain emulation and robotics to transfer human personalities into cyborgs or other non-biological carriers. The initiative's ambitious, three-step process begins with the development of an avatar or robotic copy of the self, which humans can control purely with their minds. In Phase B, an actual human brain will be transplanted into the avatar, followed by Phase C, where a digitized brain in the avatar will contain consciousness, memories, and knowledge. Finally, in stage D, a light body, like a hologram, will replace the physical form entirely, enabling humans to live forever without the limitations of any physical bodies, at all.[167,168]

Itskov views the development of these human holograms

as simply the next logical evolution of mankind, which he feels is destined to transition from the biological, material state to a higher form: that of energy and radiance. Thus, his high-tech notion is spiritual at its core. 2045 Initiative writings refer to ancient spiritual texts where human beings are described as able to evolve into entities of pure light. So far, the initiative has not met its self-established deadlines for immortality by 2045, but this well-funded project is, indeed, making steady progress toward its goal.

But all this digital and technological progress has a downside, which is that certain populations will be able to afford it and move forward into an ultra-modern, space-age lifestyle, while those without the funds to participate will be left behind—far, far behind. In fact, those who don't jump on the runaway technology train will soon be relegated to living in a new kind of Dark Age. So, if not evenly distributed around the planet, technology may very well do the exact opposite of its intention and exacerbate inequity, creating a massive social bifurcation. This underscores the importance of addressing inequity and closing the racial and gender wealth gap while advancing today's planet-saving technologies. After all, technology itself really isn't the thing that will save the world but rather the problem-solving it brings about.

Thus, the speed of technological progress is the reason why we must dedicate ourselves, as a species, to solving all the other UN SDGs right along with climate change itself. We must also focus on gender equality, clean water, sanitation, and the eradication of poverty together with everything that contributes to the social and economic equality that may one day ensure all the world's populations enjoy clean energy, safety, and the technology that assures these things. The world's problems really are linked together in an endless chain.

Observing the overlap of the seventeen UN SDGs with their technological solutions, we can really see a serendipitous unity taking place at the intersection of what's possible through science and technology; what's required by business to scale up and achieve more; and what's needed by this planet to maintain world peace while repairing the environment. That is exactly where humanity-centric innovation comes into play, building ground-breaking new business and economic models to complement the technological progress of the digital age.

But there is more to innovation than invention. As discussed, incubators and accelerators can help you translate technological breakthroughs into viable businesses, but companies don't succeed on the strength of their technology. The longevity of any endeavor is dependent upon the wisdom of its leadership. In fact, no discussion of modern business and technology is complete without a thorough grounding in the management techniques appropriate to the Gen Z population and to the management of companies mandated to stay on the cutting edge of fast-moving technology.

11

HUMANITY-CENTRIC LEADERSHIP

*The good leader is the one the people love, the bad leader
is the one the people hate, and the great leader is the one where
the people say, "We did it ourselves."*

— Lao Tzu, Chinese philosopher, author of *Tao Te Ching*,
a foundational text of Taoism

I've talked at length about the process of selecting or inventing
your own humanity-centric work to pursue, and now I'd like
to speak from the perspective of one who has begun the work,
done the work, and progressed to the point of leadership in
the same work. Having risen through the ranks in business,
I can attest to the fact that most of the world's problems are
top-down ones. With the right leadership guiding people
to be their best and work with ethically sound judgement,

human beings can achieve a lot. We could even save our species by saving our planet from our own destructive tendencies. But bad leadership can sour even the most valiant innovation. Success in business is a dance between workers interpreting the directives of management and management listening to the needs of workers, all while both partners inspire one another to create, create, create.

It truly surprises me that many businesses are still following the militaristic style of leadership popularized generations ago, and in my opinion, it's one of the primary deterrents to true innovation. Modern technologies and up-to-date goals are dependent upon minds operating from a position of inspiration, creativity, and a sense of deeply felt purpose, and business leaders need to know how to promote that atmosphere. As I've mentioned, many of today's innovative products are being promoted as "experiences" rather than simple goods or services; ironically, I think the experience of work itself is best when viewed that way, too. While my understanding of leadership is unique to my experience, it is also deeply informed by some excellent authors on the topic, and you'll find their books referenced in the annotated bibliography at the end of this book, where you can learn more about these seminal authors' work.

Inspiration, not Desperation

The secret to my success as a corporate leader is found in the fact that I observed an interesting dichotomy early in my career. You see, some leaders base their process on predicting human behavior and asking workers to fulfill the resulting expectations. The idea is to create a balance between motivation and accountability, offering rewards for progress while setting clear consequences for inaction to keep people working ever harder at the jobs they're good at, just like the way you program machines to repeat specialized tasks until their ball bearings wear out. Now, as much as I love technology, I know people are not robots.

Haarkening back to my previous lecture on human fulfillment, I think it's clear people simply need more than punishment and rewards and predictive analytics to be happy in their work and, therefore, productive. So, my leadership philosophy is to stay away from predicting anything based upon past performance but rather focus on maximizing human potential through the magic of inspiration, the hard work of removing obstacles, and the act of providing much needed encouragement. I ask a lot of myself—always trying to make the work I do matter to someone, somewhere, never wanting to just phone it in and make a buck. I assume others are the same, because I believe that's how happy, fulfilled people behave across the board. That's why I don't believe in motivating people, because I think people are already intrinsically motivated to do the right thing. However, there is a lot you can do to *demotivate* people. The first rule of leadership should be: eradicate these toxic behaviors—once and for all, no excuses!

As a service leader, I guide my teams by clearly articulating the needs of the communities and individuals who benefit from their work, ensuring that our efforts remain purpose-driven and impactful. My aim is to inspire others to do better, innovate smarter—with the objective to make people live better lives. It's those improved lives I want to encourage others to think about as they invent. As such, I see my role not as one of authority, but of service. Being part of our creation team means understanding that my purpose is to support and empower those I lead in their continuous pursuit of excellence. While I may have been seen as a "big boss," in truth, I endeavored to be a humble steward, always asking, "How can I support you today?" Creating joy in work is essentially the goal of any sound leader. Who wants to work under a burnout boss who drains the energy out of everyone, like the human version of a dead phone battery? Luckily, we have robots and AI, now, that do a lot of the repetitive stuff we used to have to do ourselves. So, more and more of today's leaders are tasked with bringing employees to a more intellectually engaging creative state. A joyful state for a person who loves a challenge. It's not difficult to manage teams in a way that brings joy to work. It's just like being a doctor: "First, do no harm." If you know what causes misery at work, well, don't do that.

Luckily, people have studied employment-based depression, and they've concluded people are most miserable when they feel their work is irrelevant, immeasurable, or invisible.[169] Such dissatisfaction comes from not having a goal worthy of your effort, a lack of recognition, and a lack of metrics by which people can tally their own progress. The simple cure for this is a corporate structure that guarantees such acknowledgement will happen regularly, not just at random intervals. To make that happen, leaders need to give people goals worthy of their effort. That means employees who are hardworking, ambitious, curious, and interested shouldn't be relegated to simply doing what they've done before. Those personality traits earn them advancement to more inspiring work that should be measured with meaningful key performance indicators (KPIs) and, importantly, credited and rewarded upon completion.

I would like to completely redefine KPIs to create greater joy in work. These new KPIs would be: Keep People Informed, Keep People Involved, Keep People Interested, Keep People Inspired, and Keep People Innovating!

But if giving credit where credit is due were all there was to leadership, this would be a pretty short chapter. Indeed, there is a lot more to it. It all begins with instilling workers with a sense of proactive responsibility. That means ending the practice of telling people to show up and do as they're told. Instead, encourage them to engage with management and take responsibility for their work. You probably know that this type of leadership is much more common in startups these days, where there is less top-down structure and more of a sense of hiring people for their overall enthusiasm for the work, then listening to their input in a spirit of empathetic collaboration.

Succeed with SMART Goals

In previous chapters, I've talked about clear goal orientation in our own lives and how we need to have personal visions to give our lives focus. The same is true in corporate leadership. Effectively prioritizing energy and time is an important business. Just telling workers the company's goal or vision is not enough. So, the best way to keep people on track and self motivated is to utilize SMART goals.[170]

SMART is a goal-setting framework that ensures objectives are clear, realistic, and trackable. It stands for **Specific, Measurable, Achievable, Relevant, and Time-bound**. A **specific** goal is clearly defined and focused on a particular outcome. A **measurable** goal is quantifiable, allowing progress to be tracked and success to be evaluated. An **achievable** goal is realistic and within reach, given available resources and constraints. A **relevant** goal aligns with broader objectives and holds meaningful value. A **time-bound** goal has a clear deadline, creating urgency and accountability. This structured approach transforms vague ideas into actionable and attainable goals. At one point in my career, we learned that consumers wanted a more soothing facial tissue, and marketing wanted a tissue that would provide a cooling sensation. I knew that to have an effective program, we needed clarity and specificity of the consumer insight, the consumer need, and the actual claim that marketing wanted printed on the box sitting on the supermarket shelf. Success came from providing guidance to the team with information as to how success of the new product would be measured, specifics as to the customer profile, an explanation of the pain point that this new product would address and how it would help people. This was all paired with a known, reasonable deadline, which made the project time bound. The old management style of demanding things "as soon as possible!" is nonsensical. After all, "soon" isn't a measurable unit of time. One-way leaders show their workers respect is to give them clear and reasonable deadlines they can work toward and have input in setting. As a leader, I've learned that if you ask a team what to do and how to do it, you get exactly what you asked for. But if you tell them what you want and why you want it, without telling them how, you always get something better than you could have imagined.

Systems thinking is a critical dimension of effective leadership that goes beyond simply establishing procedures. It involves cultivating an environment where interconnectedness, learning, and continuous improvement are embedded into the organization's culture. Through systems thinking, leaders encourage employees to see beyond isolated tasks and understand the broader impact of their work, fostering a shared vision and collective growth. By focusing on the relationships between components within the organization, leaders enable teams to identify patterns, adapt more fluidly, and pursue meaningful, sustained improvement as an integrated whole. We want our teams to focus on quality, not just expediency, so we make this easier by teaching them the thought processes that the world's best innovators use to systematically discipline their minds. Fostering a culture of accountability and recognition goes along with systemic thinking, as it engenders those long-sought accomplishments. There's an old engineering joke in product design: Good, Fast, Cheap—pick any two. But in today's world, we need to find a way to achieve all three while adding sustainability and making it easy. That's the power of humanity-centric innovation.

Create a Balance of High-Impact Tasks

Clear and meaningful goals, a systemic thought process, and plenty of recognition for achievement are all excellent elements of great company leadership, but all these elements would accomplish little without a leader who prioritizes high-impact tasks. These come in two types: quick wins and major projects. Quick wins are relatively easy tasks that can make a significant difference in a short amount of time. For instance, updating and posting a clear FAQ or troubleshooting guide on the company intranet can quickly resolve common employee questions, reducing frustration and boosting productivity. This kind of simple, high-impact action gives the employee who initiated it a gratifying sense of accomplishment with minimal effort. Employees should always enjoy a nice mix of difficult brainteasers like the cooling tissue project described above and the adrenaline rush of frequent quick wins.

These strategies keep people focused on doing their very best in a complex environment. The difference between these two attitudes is vast. When leaders value continuous improvement and personal development in both themselves and others, employee engagement and satisfaction improve. That's the kind of leadership needed in any endeavor toward humanity-centric innovation. After all, our goal is to serve the planet, not simply chase short-term gains. By embracing a holistic approach, we cultivate an environment where every team member feels valued, recognized, and inspired to bring their best contributions to a shared purpose.

Good Leadership Comes from a Growth Mindset

Considering the Four S-Curves of Life, I've noticed that whether I'm simply trying to attain financial security in the Struggle for Success or shooting for a higher-level spiritual satisfaction in the Service over Self phase, I find that, typically, other people are involved in the decisions I make, so it's best to employ leadership techniques that remove the spotlight from myself in order to illuminate and empower others.

Those in management roles have typically reached a level of stability in their own lives, making them ready to take on new challenges and growth opportunities. One of the most effective ways to do this is by fostering an environment where employees feel valued and supported, while their compensation meets their practical needs. By addressing both personal fulfillment and professional success, leaders can create a more engaged and motivated team. This service-oriented mindset is what's often called a "growth mindset," which embraces continuous learning and collaboration. It stands in contrast to the more rigid "fixed mindset," which relies on a top-down, one-size-fits-all approach to leadership. A growth mindset emerges when workers and managers alike see problems as interesting challenges and learning opportunities rather than occasions to cast blame. They view problems in this dynamic way because they believe in their own abilities to improve their skills from feedback. They say, "Yes, I can learn to do

Python coding!" and "Yes, I can learn to integrate AI into my business model!" By contrast, those with a fixed mindset either view themselves as incapable of growth or see growth and change as something that's just not worth the trouble, so they demand that others adapt to their limitations, instead. The growth mindset is inspired by the success of others, while the fixed mindset is threatened by the same thing. We see the fixed mindset predominate when one person or entity dictates behavior to control every aspect of a situation, a practice you may recognize as micromanaging.

Let's face it, micromanaging can be tempting. If you feel like you know how to do things best, getting others to follow your instructions to the letter might, at first, feel like a great success. And if you were commanding robots, not people, I'd say go for it. Program them to do exactly as told and demand perfection! But people are not robots, and even on a minute-to-minute and hour-by-hour basis, people have complex needs that must be satisfied to achieve success. That's because success at any endeavor is intrinsically tied to the different aspects of happiness explained by Maslow, Ikigai, and the Four S-Curves of Life. To achieve, innovate, and take delight in work, people must be self-motivated to do so for a trifecta of reasons: the down-to-earth practicality of a paycheck, the satisfaction of a job well done, and the act of achieving something useful to the world at large. As such, the best way for a manager to enjoy maximum power and control is to continually relinquish control. In any organization, the more you give power and control away, the more you have.

Conversely, the more you try to hold on to power and control, the more you see it slip through your fingers.

Wise leaders assume people are capable of just about anything if motivated by a combination of the company's vision and their own sense of purpose and pride in work. Instead of keeping all eyes focused on us, as leaders, we must create an environment that fosters workers knowing exactly what they're working towards, whom it will serve, and the due date, effectively taking ourselves out of the picture altogether. The focus is no longer on pleasing the boss. The "bosses" need to be in the background to enable, empower, and encourage employees to connect, communicate, collaborate, and cooperate toward meaningful, mutually shared goals. In fact, I would like to see business without bosses replaced by servant leadership.

Corporate leadership can be a lot like parenting. Truly loving parents set their children up to gradually become more and more independent until they can achieve success on their own. But weak parents keep their children dependent upon them to fulfill an ongoing need to be needed, resulting in the child lacking intellectual independence, which increases fear while squashing creativity and the ability to innovate. So, too, will any micromanaging boss. Meanwhile, the team leader who encourages independence, removes obstacles, and boosts morale to build critical thought in each employee is helping bring each unique viewpoint to the table, raising stimuli to the power of diversity until the natural course of evolution brings on groundbreaking innovation.

It's important to note, however, that this style of leadership is a lot more than just cheerleading, because focusing on employees' needs is quite different from focusing on their feelings. There may be times when you must give constructive feedback or push people harder than they're comfortable with, and that's the obligation of being a leader. In fact, some individuals don't want independence and must learn to trust themselves in order to finally enjoy it. There may also be times when you get to be the one to give accolades and compliments for jobs well done. These are all part of any leader's job. But fulfilling employees' needs is quite different from wallowing in their feelings. Some folks don't like to be asked to think critically and make their own decisions. This is where creativity is essential in guiding them toward new perspectives. If resistance remains, gently but firmly help them understand that growth requires adaptability—and that a commitment to learning and evolving is essential to our shared success. Innovative leadership thrives when it prioritizes creating a supportive, structured system that empowers employees to meet company needs and grow. Rather than simply addressing complaints, enlightened leaders listen deeply to understand challenges and proactively shape systems that foster collective success. As a corporate leader, I can attest that once you get the hang of empowering your employees instead of yourself, working in the shadows to remove obstacles to their growth, and moving out of the way to let them shine, it's addicting because its so rewarding; some might even call it spiritual. In truth, I credit my faith as a driving force in the success of my servant-style leadership.

There is nothing quite like the deeply satisfying feeling when I see my employees succeed and get all the credit for their work. The days of the "one smart guy" are gone and replaced with the belief that "none of us are as smart as all of us."

At the end of the day, I was always grateful for the trust my team had in me, and I believe our success came from creating an environment where people felt supported and empowered. I was fortunate to work with talented individuals, and when they succeeded, that was the most meaningful recognition I could have ever hoped for. The success achieved by the teams I have been associated with may not always be obvious, but I'm grateful for the consistent, innovative results my teams delivered. I believe this reflects the strength and dedication of everyone involved rather than any single person's efforts. Still, these leadership concepts are abstract. To be a truly effective leader, you'll need some very specific guidelines, you could call them formulae, that quite simply never fail. Luckily, I have two custom-made for you innovators of a humanity-centric future.

12

LEADERSHIP'S CAN'T-
FAIL FORMULAE

When we take people merely as they are, we make them worse;
when we treat them as if they were what they should be, we help
them become what they can be.

— Johann Wolfgang von Goethe, widely regarded as the most
influential writer in the German language

Strategies for leadership are somewhat complex because we
must overlap a few different concepts as we work. For instance,
every business will have a "stated strategy," and that's the set
of goals outlined in the latest presentation. When leaders
give speeches, they use carefully crafted talking points, and
all these directives are part of the stated strategy. But, often,
stated strategies fall apart and we realize that what people do

all day long is quite different. This is known as the "apparent strategy."

Apparent strategy is not an error, but a measurement of your leadership quality. A great leader gives people the recognition and encouragement they need to achieve those stated strategies, but when there is a gap between the stated and apparent strategies, all eyes look to the leader, as they should. Here, a leader needs to be accountable and use coaching, rather than discipline, to get things back on track.

Huge management mistakes are made by those who insist on doing regular performance evaluations, rating employees, and performing annual reviews. This is not the way to motivate people toward stated strategies but to simply spread a culture of mistrust. That management style is often called "pay for performance," but what you get is "pay for compliance" inspired by punitive measures and competition. They'll do as they're told to please the boss—not to improve the world, serve the customer, or achieve the overall goal, mind you, but simply to please the boss. This approach risks creating a leadership dynamic that prioritizes control over collaboration, leaving employees feeling disconnected and demotivated. Over time, they may work harder yet experience diminishing fulfillment, ultimately stifling both personal growth and organizational success.

Leadership is a Simple Formula: P=p-o+e

I use the simple equation P=p-o+e to describe a management style that has always worked for me.Here, P stands for *performance*—that elusive element that describes workers achieving the best they can do. We achieve it quite simply by assuming infinite p, or *potential*, and subtracting o, which stands for *obstacles*. After all, doesn't it make more sense to remove obstacles from peoples' paths than to push them forward unreasonably, assuming they'll magically find ways to hurdle those obstacles themselves? After you've removed all obstacles to the achievement of their potential, all workers need is a little push, a little *encouragement*. That's the e in the equation. Sometimes the biggest obstacle is in a person's mind, and they need to simply be told, "I trust you," "you can do it," and "go make it happen!"

So, to sum it up: give your workers SMART goals, but don't stop there. Next, remove *obstacles* from their paths, or, where necessary, help them understand that the obstacle is the way forward, all while becoming their personal cheering section, offering meaningful *encouragement*. You might be surprised at how self-motivated workers become when management has such faith in their natural, intrinsic desire to achieve. It has been said that people tend to rise to the level of the expectations set for them, so expecting workers to *want* to excel will help create that reality.

Providing encouragement to one's workers is, in fact, part of the act of removing obstacles, as some of our worst obstacles can come in the form of self-doubt or insecurity. Everyone is subject to these doubts now and then, but the simple cure for them is the magic created by a boss offering endless encouragement and support. In my experience, this is a management style that creates a culture of innovation, excitement, and continuous improvement in an engaged, productive workforce. The truth is, being a good, humanity-centric leader is a lot easier than being a cruel despotic dictator and a lot more fun. But it does require a paradigm shift in understanding the concept of leadership itself.

When things aren't going smoothly, reject that old-fashioned notion of simply driving workers harder. In fact, when leaders think of themselves as stewards of innovation rather than controllers or manipulators of it, this really changes the concept of management altogether. A steward is a trustee of something owned by others. As such, stewards excel at reading the wants and needs of others, almost like a wise mentor or trusted guide would do. Stewards view it as their goal to root out any areas where there might be a lack of trust, value, joy, shared purpose, vision, strategic partner-ships, and shared metrics. It's a loving approach rather than a hierarchical or paternal one. As such, we managers must become detectives who strive to figure out exactly where the obstacles lie in the paths of workers whom, we assume, would be otherwise self-motivated toward success. Removing those obstacles and replacing them with more functional elements is the work of any good steward.

When I was vice president of corporate research and engineering at quite a large firm, my teams were producing well, but I wouldn't say they were excelling. Most impor-tantly, I noticed my engineers weren't taking the kind of risks necessary to innovate with excitement and vision. The people that should have been inventing breakthroughs were merely coloring within the lines. As a leader, I want employees that are engaged, empowered, and enabled with resources, and I'm willing to invest quite a bit in getting those well-trained employees, so when folks are underperforming, I need to figure out why, and quickly. Every day that goes by with this low performance in play is a loss in value and happiness. What's more, when employee performance isn't aligned with the company's needs, my performance as their manager isn't, either. I recognize that simply removing individuals for poor performance does not address the deeper challenges at play. True progress comes from identifying and resolving the root causes, ensuring that every team member has the support, clarity, and resources needed to succeed.

If you recall from the earlier chapter "Vision, Value, and Branding," I see both my team members and myself through the lens of the value we contribute—striving for a balance where the impact we create meaningfully exceeds the resources we consume. This relationship is dynamic and interconnected. For engineers to create value, they must first receive it—from the research team that supplies consumer insights and technical data. Value flows in a continuous exchange, much like a chain reaction, where each contribution fuels the next, driving collective success. So, to implement my philosophy of P=p-o+e (performance equals potential minus obstacles plus encouragement), I needed to figure out where the obstacles were to my employees' performance.

Did team members face mental obstacles like a lack of creativity? I doubted it, as they had each been selected for their brilliance. Were the obstacles caused by something about my management style? Or was some behavior or policy in the greater company causing these brilliant scientists to tamp down their natural curiosity and excitement? To answer the question, I hypothesized that this might have something to do with the research group that fed the engineers their assignments. I wasn't sure, but it was a good place to start. So, I simply approached the business leaders and asked, "Is the research group giving you enough value?"

The answer was a simple "No!"

Like any good detective, I followed this clue to the scene of the crime. In extensive discussions with researchers, I learned the members of this team felt underappreciated. In fact, the firm had been downsizing in certain areas, and many members of the research team were afraid of losing their jobs. In short, they didn't feel that good performance was any guarantee of job security. There appeared to be patterns in decision-making that suggested favoritism, where individuals with closer relationships to certain leaders seemed to receive advantages, while others who did not align with key figures faced greater challenges in maintaining their positions. It all created a threatening and uninspiring atmosphere. Luckily, it wasn't my job to determine whether these politics were actually in play or simply an impression. The fact that the impression was there was enough to show me what was wrong: management wasn't sending researchers clear messages about company strategy and goals.

Through this process, I found that people felt insecure, that insecurity caused people to play politics, and those politics led to poor performance, which in turn led to greater insecurity. And around and around we went in this vicious cycle. The objective for me was to replace insecurity with joy-in-work. To replace politics with trust. And to replace poor performance with the creation of greater value. I believed that a system that created greater trust would improve performance, and improved performance would lead to greater joy-in-work. I translated all this into a strategy that created greater trust by focusing on building meaningful partnerships. Creating greater value by managing a portfolio of projects. Creating greater joy in work by focusing on the development of people.

All this work was eventually translated into an operational strategy for R&D.

R&D Operational Strategy

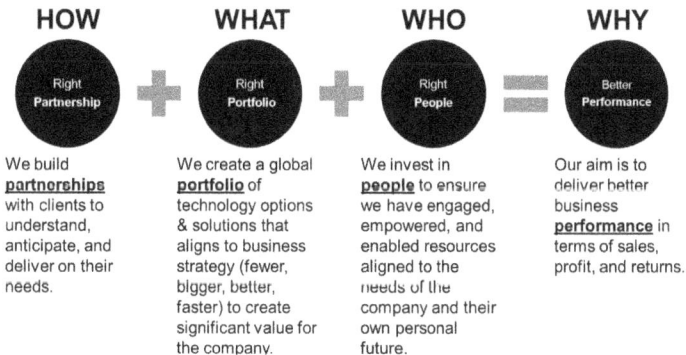

HOW	WHAT	WHO	WHY
Right Partnership	**Right Portfolio**	**Right People**	**Better Performance**
We build **partnerships** with clients to understand, anticipate, and deliver on their needs.	We create a global **portfolio** of technology options & solutions that aligns to business strategy (fewer, bigger, better, faster) to create significant value for the company.	We invest in **people** to ensure we have engaged, empowered, and enabled resources aligned to the needs of the company and their own personal future.	Our aim is to deliver better business **performance** in terms of sales, profit, and returns.

The P=p-o+e formula helped me answer all these questions and develop the strategy above, and I believe it's an equation that would work for any manager in pretty much any type of organization. But when you've resolved to be an innovative and humanity-centric company that consistently changes the world for the better, you need more than just good management, you need *inspired* leadership that keeps employees on the cutting edge of creativity. To get to that next level, I have another equation: $I=S^d/f$, or "Innovation equals Stimuli raised to the power of Diversity, divided by Fear."

Innovation is a Simple Formula: $I=S^d/F$

Charles Darwin's theory of natural selection provides the inspiration behind this equation. The basic idea is that, in nature, there is a vast amount of variation. Sure, all tortoises look like tortoises, but careful inspection reveals unique patterns on their shells, differentiation in skin color, and all kinds of other variations that distinguish each animal from others of its species. He explained the evolution of each species by positing that within these variations, nature allows for natural selection. For instance, a tortoise whose shell pattern camouflages better than those of the other tortoises has a higher chance of survival. Therefore, that shell pattern will be passed on through heredity and tortoises will evolve, over time, with traits idealized toward their survival. The same holds true in human evolution.

According to Darwin's theory, the more variety present, the more likely it is that certain animals will pass on preferred traits that strengthen the species. Thus, variation causes everything to evolve and improve. In the human world, this translates into realities such as the more varied ideas, the better; the more varied lifestyles, the better; the more varied backgrounds, the better. In the human species, as in all other species, variation is more than the spice of life, it's our very key to survival. So, when it comes to innovative ideas, the more creativity, the better. Best yet, when one person (with their own unique background, lifestyle, and viewpoints) shares an idea with a second person (with their own completely different background, lifestyle, and viewpoints), a third idea is born of the conversation. Here we see Darwin's theory of natural selection at its best, birthing not offspring, but ideas, in real time.

Returning to the $I=S^d/F$ equation, you can see how when any type of *stimuli* is raised to the power of *diversity*, it gains value-added variety, which ensures a higher likelihood of survival. So, *innovation* is born of any type of *stimuli* enhanced by different human perspectives. As much as homogeneity often makes people of like minds feel comfortable and safe, it is truly the death knell of innovation. And that's where the last part of the equation comes in.

People like to feel safe and secure in unchallenged beliefs, never facing uncomfortable truths or having to walk in someone else's shoes. So, when some humans must confront *stimuli* enhanced by diversity, it can fill them with *fear*. That *fear* (F in the equation) is the death knell of innovation, hence, $I=S^d/F$ means whatever gains you achieve by accepting stimuli raised to the power of *diversity*, your innovation will be reduced by your fear of change or organizational retribution. If your level of diverse stimuli is at a ten, but your fear of having established beliefs challenged is also at a ten, you're stuck with ten-divided-by-ten, bringing the level of innovation all the way down to one.

That's why it's every leader's job to eliminate fear in the workplace. Diversity is natural in any population, and by fostering an inclusive environment, you'll naturally reflect this richness among your employees. If you can then reduce the fear and discomfort people feel when asked to use the viewpoints of other races, genders, cultures, educations, experiences, conditions, and socioeconomic statuses, you may end up with award-winning innovation. Just as, above, I described the leader's job as one of removing obstacles, I'm holding to that but going deeper to explain that the primary obstacle to innovation, almost always, is fear, plain and simple. Replace fear with a culture of authenticity, trust, openness and acceptance, and innovation will thrive.

The Importance of Diversity to STEM Fields

For generations, the emphasis in US education has been on STEM fields—Science, Technology, Engineering, and Mathematics. It was thought that with a keen eye to these sciences, the United States could never fall behind in the arms race, the space race, or in increasing the Gross Domestic Product (GDP). But by now, educators have learned that focusing only on the hard sciences leaves people with a lack of diverse viewpoints and an intolerance to the variety that spawns innovation. As such, today, they talk about STEAM, where the A stands for "arts education." This new focus is designed to create more holistic solutions by considering social, aesthetic, and emotional aspects of design for greater accountability, transparency, and responsibility.

Furthermore, with the advent of new technologies like genetic sequencing, artificial intelligence, and advanced robotics, society will need to consider new ethical questions up to and including the very question: *what does it mean to be human?* This level of inquiry requires a higher emphasis on ethics being taught in academia, industry, and government. In response, I am encouraging education experts to further expand the acronym to STEAM-E, to include that emphasis on ethics.

I've been a science and math guy from a young age. These were always my interests as a student and professional, but numbers alone simply don't encourage the variety of viewpoints that are present in a world filled with the arts, nor do numbers answer questions about right and wrong. Fact is, looking at the world as nothing but a numbers game is a big mistake, even for a guy like me. Darwin would agree: variety (in terms of both creative interpretation and ethical viewpoints) is a lot more than the spice of life. It is the very key to human physical, emotional, and intellectual evolution itself.

In this book's annotated bibliography, you'll notice one of the pivotal books that has influenced my thinking: Steven Covey's *Seven Habits of Highly Effective People*. Luckily, Covey evolved his thinking and followed that 1989 best-seller with a follow-up book in 2004: *The Eighth Habit*, which emphasizes the importance of speaking with one's authentic voice. Doing so, he says, inspires others to find theirs, thus enhancing the existence of physical, intellectual, emotional, and spiritual variety in all discourse. Promoting such variety, says Covey, will bring on a new Age of Wisdom where people are encouraged to build and communicate their whole selves for the sake of better business in a better world. I couldn't agree more and gain strong reinforcement from the fact that this great business mind is also a proponent of stimuli raised to the power of diversity without fear.

Eureka Ranch and the $I=S^d/F$ equation

No discussion of business leadership would be complete without a mention of Doug Hall's Eureka Ranch. A unique business accelerator, Eureka Ranch is more like a think tank that helps innovators, engineers, educators, and programmers build innovation pipelines, improve innovation processes, and train their staffs to think more innovatively across the board. Eureka Ranch's founder, Doug Hall, noticed, quite early in his career, that big companies that strove to be disruptive were often held back in innovation by two things: not having the right systems in place and being just plain afraid of change, even if they didn't realize it. Thus, Eureka Ranch guides potential innovators through a proven system for innovation that gives them independence with guidance and involves three distinct units: Engineering a new company culture, incubating disruptive ideas, and performing extensive research and development toward achieving goals. Most importantly to me, though, Doug Hall was the original author of the $I=S^d/F$ leadership formula of which I am so fond, so when my company signed me up to work with Eureka Ranch, I was eager to work under Hall's team's tutelage.

Eureka Ranch works with everyone from major corporate entities to nonprofits, universities, and the public sector, always with the same philosophy that emphasizes perpetual growth, truth telling, and whole brain thinking that blends practicality with optimism. Enabling ambitious inventors to work beyond the confines of a traditional workplace, Eureka Ranch mentors aggressively pursue partnerships between like-minded innovators while valuing the intellectual curiosity and irreverent wit of their model innovator, Benjamin Franklin. Ultimately, this think tank seeks to make its courageous participants so successful that Eureka Ranch itself becomes redundant.

As Chief Scientist for a major corporate brand, I was inspired by Eureka Ranch and used their principles to drive innovation across a couple of different product lines aligned around sustainable principles. Specifically, we wanted to develop petroleum-free plastics and forest-free, paper-based tissue products. So, to implement Hall's philosophy, we first approached the issue of petroleum-free plastics with the following *stimuli*: the intention to harness the latest advancements in biochemistry and materials science to explore new possibilities. To fulfill the requirement of *diversity*, we employed a multidisciplinary team of botanists, microbiologists, and materials engineers. And to overcome the *fear* of change that tends to secretly accompany most innovation, we invested in research and development that embraced the uncertainty of this ambitious project, whose goal was to help society transition from well-established petrochemical processes to novel and largely unknown biological ones. After utilizing Eureka Ranch's system for innovation, we did succeed in developing biodegradable, petroleum-free plastics that, when implemented on a large scale, will eventually make significant inroads toward reducing carbon footprints and fostering a circular economy.

Similarly, we strove to create tissue products using tree-alternative biomass sources, such as agricultural residues, to preserve forests and promote biodiversity. To fulfill Hall's formula for innovative leadership, we again began with *stimuli*. This was a matter of leveraging agricultural science and waste management techniques to identify suitable biomass sources, which was followed by the requirement of *diversity*. As such, we brought together agronomists, environmental scientists, and process engineers to explore non-wood-fiber alternatives along with potential processing technologies.

Finally, when it came to overcoming *fear*, we confronted the fact that there was initial skepticism about the performance and consumer acceptability of non-wood fiber-tissues. We took that concern seriously and addressed it through pilot projects and consumer trials until the idea's feasibility and overall benefits were clearly confirmed. The final product we innovated turned out to be something that will help reduce deforestation and promote the general use of non-wood fibers for paper products. As such, the team championed the potential for non-traditional materials to be used in high-quality consumer products by, first, eliminating the fear factor with extensive background work, then introducing stimuli raised to the power of diversity.

13

THE FUTURE OF
LEADERSHIP IS FEMININE

How wonderful it is that nobody need wait a single
moment before starting to improve the world.

— Anne Frank, a German-born Jewish girl who kept a diary
documenting her life in hiding amid Nazi persecution during the
German occupation of the Netherlands

Some say my excitement about the latest developments in
science and technology (and how it can help rescue our
planet at this, the eleventh hour) should be tempered by the
fact that artificial intelligence sometimes seems dangerous. It
has no emotional intelligence, at least at this stage of development. Nowadays, we're seeing automation in everything
from product customer service lines to complex financial

services. It's all very efficient and often quite effective, but too much automation makes people feel isolated.

Talking to real people, even if it's just a customer service call or a technical helpline, often results in a shared laugh or a sense of finding common ground, which can brighten your day. It makes a big difference in a job when your boss is a thinking, feeling human who cares about not just your professional development but also you as a person. As much as I love equations, I don't want to be reduced to one. Nobody does.

Even though I'm a science and math guy all the way, I'm a family man first, and I know real human connections are what makes life worth living. So, is AI really a danger to our wellbeing? Is its growing dominance going to create massive unemployment, strip society of its shared humanity, and give us a world without self-awareness, empathy, or social skills? These are good questions to ask in these changing times, but I'm not worried about it, because humans will always offer something robots currently don't: emotional intelligence. This increasingly valuable skill doesn't require an education or programmng.

Previous industrial revolutions caused a lot of unemployment when manual laborers were replaced by machines, and along with that came social restructuring. The long-term effect of each new technological development was that the highly educated found ways to work with the new society being built, while those doing manual labor and other low-education jobs tended to get paid less. Industrial revolutions, historically, have increased the wage gap, diminishing the middle class with each "forward" step. We don't want to repeat that pattern! But AI is different for a lot of reasons.

AI builds on, integrates, and potentially even replaces human intelligence with super-human precision, complex systems management, superior processing speed, data analysis abilities, pattern recognition skills, and exceptional task execution. Considering where AI is now with these skills, and the fact that it is only improving, real humans will never be able to match it. Why try? But AI can't do what good team managers and corporate leaders do. AI-informed robots can't give authentic heartfelt encouragement to human workers. They can't build soulful partnerships among like-minded people. And they certainly can't inject genetic diversity into the workforce. Robots do tasks and do them well, but people are needed in a world culturally and behaviorally defined by human beings. There will always be a need for real people at all education and skill levels in the workforce: AI bots may have a superior programmed IQ, but humans bring the rich, innate abilities of both IQ and emotional intelligence.

For this reason, to make ourselves more useful and employable in an AI-dominated world, we will have to change our work styles to be more open, emotional, empathetic, spiritual, philosophical, accepting, and encouraging. Essentially, we humans will evolve to be even *more* human than we already are.

I predict that where people once attended trade schools and job certification programs to become qualified for task-oriented jobs, they'll soon attend conflict resolution schools where they develop skills around counseling, human empathy, leadership, and bridge building. Where people once got MBAs that emphasized protecting that all-important "bottom line," they'll now learn a new way focused on conflict resolution, imagination, setting a vision, empowering others, and, most importantly: ethics. These skills will qualify people at all education levels to provide the human interaction that will become more, not less, in demand as AI takes over life's mundane tasks.

For centuries, leadership has been defined by traits like assertiveness and competition—qualities often associated with traditional, hierarchical power structures. But as we step into an era where artificial intelligence challenges the very notion of human intelligence, the most impactful leaders won't be those who simply command authority; they'll be those who balance strength with service, strategy with empathy, and decisiveness with collaboration.

The 21st-century high-tech hero isn't just measured by IQ (Intelligence Quotient) but by a new kind of EQ—one that goes beyond emotional intelligence to encompass Empathy, Ethics, Empowerment, Envisioning, Entrepreneurship, and Encouragement. These are the qualities that will define the leaders who thrive in a world shaped by AI, automation, and unprecedented digital transformation.

In this hyper-connected world, success isn't about domination—it's about creation, connection, and contribution. The best leaders will foster environments where innovation flourishes through inclusion, where technological advancements serve humanity, and where the future is built on a foundation of trust, creativity, and shared purpose.

It's time to redefine leadership—not as a position of control, but as a catalyst for transformation. The future belongs to those who dare to lead with both vision and heart.

Be a High-Tech Hero

Throughout my career, I've always tried to hold true to my vision of helping create humanity-centric companies and helping children live a full life. As a leader, I've always tried to emphasize joy in work. As an innovator, I've looked toward the UN SDGs as guides to doing right for people and the planet. But these ideas didn't come from out of the blue. They have been inspired by the reading I've done and the conferences I've always believed in doing the right thing, guided by a commitment to integrity, service, and long-term impact. While I know I have room to grow, I strive to lead by example—prioritizing humility, empathy, and a mindset of stewardship. I am dedicated to fostering sustainable business practices that protect our planet and ensure a thriving future for generations to come. By embracing principles of social responsibility and ethical leadership, I aim to create value not just for today, but for the long term, ensuring that businesses operate in a way that benefits both people and the environment. My focus is on using my skills and experience to drive meaningful progress, balancing economic success with a greater commitment to sustainability and human well-being. Whether you share this perspective from a moral, strategic, or purely practical standpoint, we all have the same goal—to repair the damage we've done, build a more sustainable future, and enable people to live and work in harmony with the world around us. If, like me, you see yourself and the unique time period in which you live as one piece of a big puzzle that spans time and space, then you must agree that the puzzle is solvable . . . if we all do our part. Compare saving the earth to one huge video game. Level up, stay alert, recruit allies, and believe in your mission, and you'll succeed. You will. Because, unlike the members of my generation, you Gen Zers are the first generation to enter the workforce with all the tools necessary to save the planet and humanity right there at your fingertips. You just need to sharpen them and use them strategically while keeping the enemies of sustainability at bay.

Googling some of my humanity-centric ideas, one day, I saw an incredible painting online. It had won a UK "Vision of Science" award by Adam Nieman.[171] The beautifully rendered image was a globe of planet Earth with three, much-smaller spheres, superimposed. I found it fascinating because the three spheres represented all the water, all the freshwater, and all the available water in the world, and they were small in comparison to the Earth.

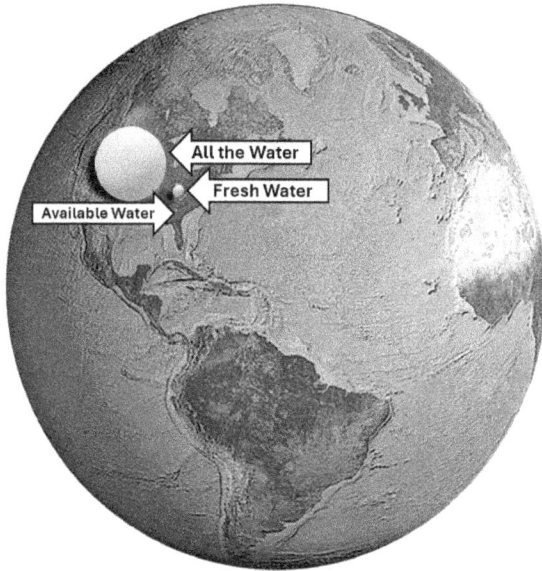

We all know that the surface of the earth is mostly water: 72 percent, but this illustration made viewers aware of the fact that our oceans and seas are actually a thin film of water, and a lot of it is locked up in polar ice caps. If you look at the actual volume of water available, rather than the surface area it takes up, you realize how very little ocean we have by comparison to this big rock we call home. The third bubble, representing available water, not trapped in the polar ice caps, was the smallest one, about the size of Atlanta, Georgia, when compared to the land behind it. Freshwater is a renewable resource through the hydrodynamic cycle, but, considering the eight billion people on Earth drinking it, water is a precious resource. The illustration brought into stark relief just what a delicate planet this is, how lucky we all are to inhabit it, and how very much attention we must pay to keeping it beautiful and viable for life.

I was so enamored by the illustration, I looked up the artist's name online, found his phone number, and called to ask if I could purchase it. The artist was excited to have a customer call out of the blue like that, but then he thought twice.

"What are you going to do with it?" he asked.

I told him I'd hang it in my office, so he asked where I worked.

I told him the multinational firm I worked for. I was proud of my accomplishments there and what our innovative products had done to promote health and create convenience and comfort all over the world. But the artist didn't see it that way. He was aware of the wide range of work these kinds of multinational corporations get involved with, and not all of it is humanity centric. Not by a long shot. He accused the same company of promoting deforestation and pollution in many regions of the world. I knew it was a bit of a cop-out for me to shrug and say, "But that's not my department." So, I didn't say it, but still, it wasn't my department. I had spent the bulk of my career working on clean water initiatives and world sanitation projects for the same company that, it could be argued, did the opposite type of work in other regions of the world. In my defense, I argued that the way I saw myself was as a change agent inside the system, transforming the organization from the inside, trying to move it toward sustainability from within. I had always adhered to my be-my-own-boss mindset, believing that the corporation worked for me to give me opportunities to achieve humanity-centric goals using its financial and technical resources. Sure, the firm thought I worked for it, but I viewed myself, instead, as an undercover humanity-centric billionaire helping a billion people over the course of my lifetime. Well, the idealistic (and very well-informed) artist didn't buy it. He said he was sorry, but he couldn't sell me the illustration. I admired his dedication to sustainability, even if I didn't agree with his viewpoint about my work. But, in the end, he was inspired by passion and purpose, and he helped me contact an agency in New York that had rights to provide prints of the artwork. As I type this, that artwork lies on the wall in my office where I look at it every day and I am reminded in the importance of humanity-centric innovation.

That said, frankly (for a Boomer) I think I've done a good deal toward building a more humanity-centric world, and I haven't quit, yet. But it's time for me to pass the torch to you: a new generation of leaders who are far more well-equipped than I ever was. That said, I'll be here to support and contribute in any way I can as you shape the future. You have grown up knowing the problems we face on a planet where global warming causes increased climate change every year. You know what it's like to worry that your children won't be able to find a conflict-free, healthy place to live. You young men are deeply aware of the gender and racial pay gaps and the world inequity it causes. You young women are naturally attuned to the importance of rejecting the old-fashioned,masculine management style for something more oriented toward consensus, joy, and balance.

In this new era, the most effective leadership will not

be defined by gender but by a shift in mindset—from dominance to collaboration, from control to care. The traits we've long associated with feminine energy—empathy, nurturing, inclusivity, deep listening, and empowerment—are exactly what the 21st century demands. We need leaders who coach rather than command, who foster connection over competition, and who see success not as winning at all costs, but as lifting everyone toward a better future. The old top-down, winner-take-all model has outlived its usefulness in a world that depends on resilience, diversity, and cooperation. What's emerging now is a more balanced and evolved form of leadership—one that values emotional intelligence as much as strategic vision. This is the leadership that will help us solve our biggest challenges and build a world truly centered on humanity.

Our generation, and those before us, have been seen as contributors to the planet's current challenges. While we worked with the tools we had, we now recognize the need for a new approach. You have the knowledge and resources to innovate differently, and that's where real change begins. You members of Generation Z carry in your intellect, imaginations, and pockets the technological tools that can fix the ravages of the last four industrial revolutions. So, what do you say? Truly, there is no better future you could plan for yourself and your family than becoming a "billionaire" in the most modern sense of the word. So, go on, become a billionaire!

ANNOTATED BIBLIOGRAPHY

Bakke, D. (2005). *Joy at Work*. PVG. - As a leader, to foster joy in work, we must empower our teams by delegating decision-making authority, recognizing their contributions, and aligning their work with their personal values, purpose, and passions.

Benyus, J. (1997). *Biomimicry: Innovation Inspired by Nature*. William Morrow. - Janine Benyus teaches us that Mother Nature, having refined life-sustaining systems over 4.5 billion years, offers invaluable lessons for creating sustainable solutions through mimicking natural processes and principles.

Block, P. (1993). *Stewardship*. Berrett-Koehler Publishers. - Advocates for a leadership approach that prioritizes service, empowerment, transparency, and shared responsibility to create a more equitable and sustainable organizational culture.

Brook, A. (2006). *The Human State*. Self-published. - To empower people to create value from their data while ensuring its protection, we must adopt principles which emphasize transparency, ethical stewardship, and the safeguarding of personal information to maintain trust and integrity.

Buford, B. (1994). *Halftime*. Zondervan. - I learned that shifting focus from personal success to significance involves embracing a life of service over self-interest, ultimately leading to a more meaningful and impactful existence.

Chouinard, Y. (2005). *Let My People Go Surfing: The Education of a Reluctant Businessman*. Penguin Books. - Patagonia is a great example of humanity-centric innovation by prioritizing environmental responsibility, employee well-being, and ethical business practices, demonstrating that a company can be successful while staying true to its core purpose.

Coelho, P. (1988). *The Alchemist*. HarperCollins. - I came to learn and believe that when you are committed to your personal mission and follow your dreams with determination, the entire universe conspires for your success.

Covey, S. R. (1989). *The 7 Habits of Highly Effective People*. Free Press. - In my first year working in industry, I came to understand the profound impact of these principles, which helped me develop essential leadership qualities such as proactive behavior, clear goal-setting, and effective prioritization, ultimately making me a better leader.

Dweck, C. S. (2006). *Mindset: The New Psychology of Success*. Random House - A growth mindset is essential for stepping out of your comfort zone and tackling humanity's biggest problems, as it encourages continuous learning, resilience, and the belief that abilities can be developed through effort and perseverance.

Dawkins, R. (1976). *The Selfish Gene*. Oxford University Press. - Helped me understand that just as evolution occurs through the transmission of genes from one generation to another, innovation happens through the sharing of memes from one person to another, illustrating how ideas and cultural elements evolve and drive progress.

Deming, W. E. (1986). *Out of the Crisis*. MIT Press. - No book has had a greater impact on how I think about management than W. Edwards Deming's "Out of the Crisis," which revolutionized my entire approach to leadership by emphasizing the importance of systemic thinking, continuous improvement, and the pivotal role of leadership in fostering a culture of quality, accountability, and sustainability.

Diamandis, P., & Kotler, S. (2020). *The Future Is Faster Than You Think*. Simon & Schuster. - This book profoundly influenced my thinking on 21st-century business models, highlighting their crucial role in making humanity-centric innovation economically viable by leveraging exponential technologies to create sustainable and scalable solutions.

Frankl, V. E. (1946). *Man's Search for Meaning*. Beacon Press. – His teaching influenced my understanding of the importance of purpose and meaning, highlighting that striving to accomplish something beyond oneself and one's immediate circumstances is essential for enduring fulfillment and resilience.

Fussler, C. (2004). *Driving Eco-Innovation*. Pitman Publishing. – This book significantly influenced my understanding of how businesses can act as powerful engines for enhancing the sustainability of our planet by integrating eco-friendly practices and innovative solutions into their core strategies.

George, R. (2008). *The Big Necessity: The Unmentionable World of Human Waste and Why It Matters*. Metropolitan Books. – This book underscores the critical importance of improving global access to toilets to safeguard health, dignity, and safety, particularly highlighting the severe risks women face when lacking secure sanitation facilities.

Gallwey, W. T. (1974). *The Inner Game of Tennis*. Random House. - Expanding on the model taught in Timothy Gallwey's "The Inner Game of Tennis," I created the equation that performance is equal to potential minus obstacles plus encouragement. By removing obstacles and providing encouragement as a leader, you can help others realize their full potential and exceed your performance expectations.

Hines, A. (2007). *Thinking about the Future: Guidelines for Strategic*

Foresight. Social Technologies. – It is from Andy Hines that I learned about the Futures Cone, which maps various possible, probable, plausible, and preferable futures, and recognized that humanity-centric innovation lies within the cone of preferred futures, guiding us toward sustainable and equitable advancements that align with our highest aspirations for humanity.

Johnson, S. (2010). *Where Good Ideas Come From.* Riverhead Books. - I learned that your next great idea often emerges not from working alone but through sharing and discussing your ideas with others, as such interactions spark new insights and lead to the birth of even better ideas.

Kelly, K. (1994). *Out of Control: The New Biology of Machines, Social Systems, and the Economic World.* Basic Books. - the secret to organizational change lies in the hive mind, which requires three key elements: a compelling vision that everyone understands, complete freedom within a framework, and constant communication about progress toward the vision.

Krznaric, R. (2020). *The Good Ancestor.* WH Allen. - I've come to understand that the actions we take today will determine whether our descendants will merely survive or thrive. When future generations write their history books, they will either praise us for sparking a new renaissance or blame us for creating a dark age; the outcome depends on our commitment to making sustainable and long-term decisions now.

Lencioni, P. (2007). *The Three Signs of a Miserable Job*. Jossey-Bass. - By doing the opposite of the three signs of a miserable job—irrelevance, immeasurability, and invisibility—leaders can foster greater joy in work.

MacKenzie, G. (1998). *Orbiting the Giant Hairball*. Viking Adult – This provides a powerful reminder that "there is a masterpiece inside of you; if you don't paint it, it won't get painted. No one else can paint it but you," highlighting the unique potential within each of us and the irreplaceable value of our individual purpose, passion, and contributions.

Pauli, G. (2010). *The Blue Economy*. Paradigm Publications. – The Blue Economy shares similarities with the concepts of the circular economy and humanity-centric innovation by emphasizing sustainable business models that mimic natural ecosystems, create value from waste, and prioritize the well-being of both people and the planet.

Polman, P. (2020). *Net Positive: How Courageous Companies Thrive by Giving More Than They Take*. Harvard Business Review Press. – This book closely relates to humanity-centric innovation by showcasing how businesses can create value for society and the environment, illustrating practical examples of companies that integrate sustainable practices and prioritize long-term well-being over short-term profits.

Quinn, D. (1992). *Ishmael: An Adventure of the Mind and Spirit.* Bantam/Turner. - Based on the teachings of Daniel Quinn it is crucial to change our cultural narratives to create a more sustainable world by recognizing the interconnectedness of all life and shifting from a mindset of dominion over nature to one of stewardship and harmony.

Raworth, K. (2017). *Doughnut Economics: Seven Ways to Think Like a 21st-Century Economist.* Chelsea Green Publishing. – This book parallels humanity-centric innovation by advocating for an economic model that balances the needs of all people within the means of the planet, emphasizing sustainable development and well-being over traditional growth metrics, which aligns with the principles of creating solutions that prioritize human and environmental health.

Rifkin, J. (1980). *Entropy: A New World View.* Viking Adult. - The concept of entropy is applied to economic and social systems, illustrating the unsustainable nature of current economic practices and the necessity for more energy-efficient ways of living and organizing society. This principle greatly influenced my thinking about sustainable development, emphasizing the importance of integrating ecological principles into economic systems to ensure long-term sustainability and efficiency.

Sachs, J. (2005). *The End of Poverty: Economic Possibilities for Our Time.* Penguin Press. – This influenced my thinking by demonstrating that with the right strategies and international cooperation, it is possible to eradicate extreme poverty and create economic opportunities for all, underscoring the importance of sustainable development and global collaboration.

Senge, P. M. (1990). *The Fifth Discipline: The Art and Practice of the Learning Organization*. Doubleday/Currency. - Strategies taught here can be used to solve global problems by promoting systems thinking and continuous learning, enabling organizations to develop holistic, sustainable solutions that address root causes and leverage new business models and technologies.

Sinek, S. (2019). *The Infinite Game*. Portfolio. - Leaders and organizations should adopt an infinite mindset, focusing on long-term success and sustainability rather than short-term gains, by continuously adapting, embracing change, and prioritizing values and purpose over immediate results.

Wheatley, M. J. (2006). *Leadership and the New Science: Discovering Order in a Chaotic World*. Berrett-Koehler Publishers. - A new paradigm for addressing significant problems by applying principles from quantum physics, chaos theory, and complexity science to foster creativity, resilience, and adaptability within organizations.

Crosby, P. B. (1979). Quality is Free: The Art of Making Quality Certain. McGraw-Hill. - A foundational work on quality management that introduces the concept that quality, when approached correctly, can be achieved at no additional cost. Crosby emphasizes "doing it right the first time" to reduce costs associated with errors, advocating a proactive approach to quality as a key to organizational efficiency and customer satisfaction.

APPENDIX A

DATA MANAGEMENT OPPORTUNITIES

Data Cooperatives

These data exchange platforms leverage blockchain for security and transparency while empowering individuals to collect their data from sources such as social media, health records, and e-commerce, while ensuring transparency, control, and security, all to share it with buyers in exchange for rewards, turning personal data into a valuable asset.

MIDATA	Polypoly
DataWallet	Eva Coop
Digi.me	Ubiquitous Commons
Data Union	Mnémotix
Salus Coop	dOrg.tech
LBRY	

Decentralized Privacy-Preserving Data Marketplaces (dPDMs)

These decentralized data exchange protocols are designed to unlock data for AI and other applications, allowing users to share and monetize their data securely while retaining control over it. Blockchain technology ensures data privacy, integrity, and security, facilitating a transparent and efficient marketplace for data assets that fosters inclusivity and collaboration, democratizing access to decentralized networks.

Ocean Protocol	Datum
Opiria	DcentAI
Dataeum	Streamr

Data-Encryption Technology Available to Consumers
Brave Browser

This privacy-focused web browser blocks ads and trackers by default, providing a faster and more secure browsing experience. Brave rewards users with Basic Attention Tokens (BAT) for opting into privacy-respecting ads, enabling them to monetize their browsing data.

VeraCrypt and BitLocker

These free, open-source disk encryption softwares enhance security for data storage and allow users to create encrypted volumes and partitions, protecting data from unauthorized access. They are widely used for robust encryption of sensitive data on personal computers and external drives, providing robust encryption to safeguard personal information.

Data-Encryption Technology Available to Data-Monetizing Companies

These decentralized platforms specialize in data capture and processing utilizing AI and text analytics and enabling the creation and execution of self-executing smart contracts running on the blockchain. Helping businesses unlock hidden data from various sources, the platform enables organizations to monetize their data by transforming it into actionable insights with trustless agreements between providers and consumers.

Ethereum	Utopia Analytics
Datumize	Chainlink

Data Regulation and Governance Agencies
GDPR

The General Data Protection Regulation (GDPR) is a comprehensive data protection law in the European Union that regulates how personal data is collected, processed, and stored. It grants individuals greater control over their data and imposes strict obligations on organizations to protect privacy. GDPR aims to enhance data security and transparency, ensuring that personal data is handled responsibly.

CCPA

The California Consumer Privacy Act (CCPA) is a state-wide data privacy law that grants California residents new rights regarding their personal information. It allows consumers to know what data is being collected, request deletion of their data, and opt-out of data sales. The CCPA aims to enhance consumer privacy and promote transparency in data handling practices.

Data Regulation Compliance Technologies

These privacy, security, and data governance platforms help organizations comply with global regulations while providing tools for consent management, data mapping, and privacy impact assessments, ensuring that businesses handle personal data responsibly while supporting compliance with regulations like GDPR and CCPA, enhancing data privacy and protection.

OneTrust

TrustArc

APPENDIX B

SUSTAINABLE BUSINESS INCUBATORS AND ACCELERATORS

We Make Change:
We Make Change connects skilled volunteers with social enterprises and NGOs worldwide to tackle global challenges. They leverage the power of volunteerism to create positive social impact.

https://www.wemakechange.org/

JA Worldwide
The participants for JA Worldwide's Build for Earth accelerator are chosen from Hack for Earth, a hackathon for climate change. Winners enjoy diverse commercial and institutional partnerships and sponsorships, which provide backing for

innovation that enables JA Worldwide to transform the lives of millions of young people every year, who, in turn, transform the world.

https://www.jaworldwide.org/

Acumen

Acumen invests in companies, leaders, and ideas that are changing the way the world tackles poverty. It aims to create a world beyond poverty by investing in social enterprises, emerging leaders, and breakthrough ideas.

https://acumen.org/

The Unreasonable Group

This incubator supports growth-stage entrepreneurs who are solving significant global challenges, ranging from clean energy to sustainable agriculture, by providing them with resources, mentorship, and access to capital.

ttps://unreasonablegroup.com/

Ashoka

Ashoka identifies and supports the world's leading social entrepreneurs—individuals with innovative and practical ideas for solving social problems. They provide these entrepreneurs with the tools and networks they need to achieve large-scale social change.

https://www.ashoka.org/en-us

Village Capital

Village Capital supports early-stage ventures that are solving problems in sectors like education, health, energy, and financial inclusion. They use a peer-selected investment model and focus on entrepreneurs who are often overlooked by traditional investors.

https://vilcap.com/

Echoing Green

Echoing Green provides fellowships, seed-stage funding, and strategic support to emerging leaders working to bring about positive social change across the globe.

https://echoinggreen.org/

Hult Prize Foundation

This global competition challenges university students to solve pressing social issues through social entrepreneurship. The winners receive significant funding and support to turn their ideas into reality.

https://www.hultprize.org/

Chandler Foundation

The foundation supports and collaborates with organizations and leaders working to build prosperous and equitable societies. They focus on governance, education, and economic opportunities.

https://www.chandlerfoundation.org/

MassChallenge

A global, zero-equity startup accelerator that supports high-impact, early-stage entrepreneurs across industries, including social impact. They provide mentorship, funding, and access to a network of business leaders.

https://masschallenge.org/

The Biomimicry Institute's Ray of Hope Accelerator

Supporting high-impact, nature-inspired startups, Ray of Hope provides $15,000 in non-dilutive funding and over $50,000 worth of in-kind services including tailored coaching, investor introductions, a nature retreat, and comprehensive training materials covering topics such as Impact and Sustainability Business Training, Innovation Storytelling & Amplification, DEI, and Founder Mental Health.

https://biomimicry.org/innovation/accelerator/

Anderson Center for Sustainable Business

A division of Georgia Tech's Scheller College of Business, the Anderson Center for Sustainable Business is a business incubator in the form of a university program. It acts as a catalyst and connector for students, research faculty, companies, and entrepreneurs to create an environment where business-driven solutions to sustainability challenges can and thrive.

https://www.scheller.gatech.edu/
centers-and-initiatives/ray-c-anderson-center-for-
sustainable-business/index.html

Ignite

Ignite offers £100k to ten teams every year, running programs across the UK and Ireland and consulting with accelerators in countries such as India, the United Arab Emirates and the US. To start and scale sustainable businesses, Ignite challenges founders to validate core business assumptions and build scalable traction.

https://www.ignite.io/

Blue Startups

Blue Startups helps scalable-technology companies compete globally, stimulating economic growth in Hawaii and creating new business opportunities for entrepreneurs with mentorship, access to business expertise, product testing, peer review, pitch development, introductions to investors, and seed funding.

https://www.bluestartups.com/

World Startup Factory

World Startup Factory's Changemaker Platform is a digital accelerator that also holds live events. Embracing rapid experimentation, agile development, and continuous learning, it serves those who live and breathe impact-driven entrepreneurship with expert mentorship, tailored guidance, and dynamic peer-to-peer collaboration.

https://worldstartup.co/

Katapult Ocean Accelerator

Katapult Ocean invests in startups with a positive impact on the ocean using a three-month intense program with a focus on growth, investor readiness, leadership development, exponential tech, and introductions to the Norwegian and global ocean tech ecosystem, covering transportation, ocean health, harvesting, energy, and new frontiers. Participants pay a $50,000 program fee and receive a $150,000 investment in exchange for 8% equity.

https://katapult.vc/ocean/

100+ Accelerator

The 100+ Accelerator fuels the growth of startups developing critical sustainability solutions by enabling five of the largest consumer goods companies in the world to work with start-ups to maximize collective impact. By funding up to $100,000 per startup pilot, the partners test and validate each solution to scale and transform their global supply chains.

https://www.100accelerator.com/

BlueSwell Incubator

BlueSwell was formed to support the creation and growth of startups with scalable solutions that enhance ocean health, sustainable ocean industry, and global resilience. Its goal is to build the capacity of new founders to convert big, ocean-focused concepts into profitable, sustainable businesses by bridging a gap of funding and mentorship.

https://blueswell.sea-ahead.com/

European Startup Prize for Mobility

This public-private initiative aims to scale up smart and sustainable mobility startups across Europe. Each year, more than 600 startups apply to its Acceleration and Investment Programme, which provides multiple investment opportunities, tailor-made mentoring, connections to EU organizations, and visibility in front of EU decision makers.

https://startupprize.eu/

Founders Bay Accelerator

The Founders Bay Accelerator is an equity-free, six-month intensive program to support startups by building partnerships with its teams, understanding their unique needs, and continuously refining the work through feedback loops.

https://founders-bay.io/

Freigeist

This venture capital firm backs extraordinary founders at an early stage. Making only one or two investments annually, the firm emphasizes technologies like AI, robotics, synthetic biology, and quantum-computing, then continues to work with mentees on a comprehensive roadmap for the next twelve to twenty-four months.

https://freigeist.com/

Greentech 2022 Village Capital

Greentech improves the scalability of gender-diverse start-ups in Europe. Supporting thirty early-stage, small and medium businesses (SMBs) at a time, it provides training, expert advice, network support, and funding to help close the gender financing and resource gap.

https://vilcap.com/current-programs/
greentech-europe-2022

StartUp Bootcamp

This six-month accelerator program for early-stage impact entrepreneurs focuses on developing sustainable business models, impact management, strategies for scaling, and access to financing for every development stage.

https://www.startupbootcamp.org/

Impactivs

Impactivs selects startups based upon their priorities, key skill gaps, and particular industry as well as where entrepreneurs are in their journeys, then provides tailor-made, mentored, module-based digital acceleration programs.

https://www.impactivs.com/

Obratori

This early-stage venture capital fund & accelerator features a coworking space in the business district of Marseille, France. Dedicated to supporting innovative startups in wellbeing and green tech, this seed investment fund focuses on sectors, services and products that meet its commitment to supporting innovation and high-potential projects.

https://obratori.com/

Sustainable Accelerator, London

Sustainable Accelerator is managed by sustainability veterans including successful entrepreneurs and experienced venture capitalists. With an interest in leveraging industry networks, experience, and seed capital, it mentors innovative entrepreneurs from start-up idea to exit.

https://www.sustainableventures.co.uk/investment

Paris Techstars Sustainability Accelerator

Specifically for entrepreneurs who are French or willing to settle in the Paris ecosystem to thrive globally, Techstars invests in early-stage, high-growth tech startups that want to solve some of the greatest challenges our planet faces today, such as climate change, pollution, poverty, and disease.

https://www.techstars.com/accelerators/paris

Tribe Accelerator

Tribe Accelerator is a go-to-market-focused accelerator emphasizing fundraising and developing product, business, and marketing strategies. It provides its portfolio companies a hyperconnected platform to accelerate their innovative usecases through a global network of MNCs, government agencies, and top tech companies.

https://tribex.co/accelerator/

VenturePad Climate Impact Fund

This full-service accelerator and business hub helps top impact start-ups launch and grow, and scale in Marin, North Bay, and regionally. VPA's goal is to incubate, launch, and scale twenty new businesses a year through two cohorts running approximately five to seven months each.

https://www.venturepad.works/accelerator

Vertuelab Impact Accelerator

VertueLab provides funding and holistic entrepreneurial support to cleantech startups. Its $5M Climate Impact Fund invests in pre-seed, early-stage companies expected to have the most impact on the climate. Its programs include a fifteen-week accelerator, federal grant assistance, and interns.

https://www.vertuelab.org/impact-fund

Windsail Capital

This leading venture capital investor provides growth capital for the clean economy, specifically to companies advancing energy innovation and sustainability. With its focus on overlooked companies such as service businesses, the firm offers a unique investment approach providing flexible financing solutions that facilitate growth while minimizing dilution.

https://incubatorlist.com/windsail-capital/

Toilet Board Coalition

Focus: The Toilet Board Coalition supports innovative businesses that address global sanitation challenges through market-based solutions. They aim to achieve sustainable sanitation for all.

https://www.toiletboard.org/

Ukukhula

This incubator grows entrepreneurs mentally, spiritually, and financially in the fields of water and sanitation, exponential technology, web-based platforms, e-learning, infinite computing, sensors and networks, 3-D printing, nanomaterials and nanotech, artificial intelligence, robotics, and genetics and synthetic biology. Mentors require no no out-of-pocket expenses for the coachee.

https://www.ukukhula.nl/

The Business Collective Pte Ltd

The Business Collective is an eight-week business accelerator that invests in products, services, and companies with a view to highly impact society, especially high growth sectors in Southeast Asia. Its focus is mainly on Healthcare/Health-tech, Clean Energy (production and conservation), Education, Safety & Security, Sanitation, E-commerce, and Technology.

https://thebusinesscollective.sg/

Sigma Accelerator

This social startup accelerator program offers a four-month intensive curriculum guided by corporate and ecosystem partners on developing scalable innovation initiatives. It aids startups creating affordable products and services for underserved markets in India focusing on healthcare, cleantech, smart-agri, sanitation, and education.

https://zinnov.com/sigma-accelerator/

ABOUT THE AUTHOR

Pete Dulcamara is the founder of Pete Dulcamara & Associates, a consultancy dedicated to "helping create businesses that improve people's lives." Central to Pete's work is the concept of humanity-centric innovation, which leverages exponential technologies such as AI to solve humanity's biggest challenges in economically viable ways. He redefines a "billionaire" as someone who impacts a billion lives.

With extensive expertise in materials science, life sciences, AI and digital transformation, Pete has a distinguished career leading global innovation. As Chief Scientist and Vice President of Corporate Research & Engineering at Kimberly-Clark, he led the discovery, development, and delivery of science and technology across all businesses, brands, and regions globally. A significant part of his legacy is advancing the field of 'superabsorbency,' a breakthrough method for the collection and disposal of various forms of human waste. This unique expertise has led to his taboo-breaking ideas for using these products to collect biowaste in a circular economy. He also led employee and community engagement as Site Leader for the company's Innovation and Operational Excellence Center.

Prior to Kimberly-Clark, Pete held senior roles at The Dow Chemical Company, driving advancements in R&D, sustainability, and corporate innovation. His expertise spans AI-driven R&D, operational efficiency, and strategic growth.

Pete currently serves on the Executive Advisory Board for the High Impact Technology Fund at Stanford University and contributes as a board member for Bassett Mechanical. He is a Chemicals & Materials Partner at FutureBridge, Executive Partner at Pilot Lite, a Partner at Endeavor.ai, and a partner at Midpoint Consulting, where he focuses on AI education for the modern workforce. As a founding member of the Advisory Board for the Central Wisconsin AI Center, Pete is advancing a three-year goal to engage over 100,000 workers and 300 businesses in creating an AI-ready workforce.

Pete's insights have global resonance. He has delivered a TEDx talk on humanity-centric innovation and regularly speaks on the transformative potential of exponential technologies, earning acclaim for making complex topics accessible and inspiring actionable innovation.

Pete lives with his wife, Gina, in Neenah, Wisconsin, where he drives innovation and economic development at the intersection of technology, sustainability, and business transformation.

To connect with Pete, visit his website at
www.petedulcamara.com

ENDNOTES

1   United Nations, *Report of the United Nations Conference on Sustainable Development*, Rio de Janeiro, Brazil, 20-22 June 2012, A/CONF.216/16, 2012, https://digitallibrary.un.org/record/737074/files/A_CONF.216_16-EN.pdf.

2   United Nations Framework Convention on Climate Change, *Report of the Conference of the Parties on its Twenty-First Session, Held in Paris from 30 November to 13 December 2015*, FCCC/CP/2015/10/Add.1, January 29, 2016, https://unfccc.int/documents/9097.

3   United Nations, *Report of the Secretary-General on the 2019 Climate Action Summit*, New York, 23 September 2019, https://digitallibrary.un.org/record/3850027/files/Report_of_the_Secretary-General_on_the_2019_Climate_Action_Summit.pdf.

4   United Nations Framework Convention on Climate Change, *Paris Agreement*, adopted December 12, 2015, entered into force November 4, 2016, https://unfccc.int/sites/default/files/english_paris_agreement.pdf.

5   William Gibson, interview by Neal Conan, *Talk of the Nation*, National Public Radio, November 30, 1999, https://www.npr.org/2018/10/22/1067220/the-science-in-science-fiction.

6   Erin Wayman, "When Did the Human Mind Evolve to What It Is Today?" *Smithsonian Magazine*, June 25, 2012, https://www.smithsonianmag.com/science-nature/when-did-the-human-mind-evolve-to-what-it-is-today-140507905/.

7   Klaus Schwab, "The Fourth Industrial Revolution," World Economic Forum, accessed February 25, 2025, https://www.weforum.org/about/the-fourth-industrial-revolution-by-klaus-schwab/.

8   Jason Dorrier, "IBM's New Computer Is the Size of a Grain of Salt and Costs Less Than 10 Cents," *Singularity Hub*, March 26, 2018, https://singularityhub.com/2018/03/26/ibms-new-computer-is-the-size-of-a-grain-of-salt-and-costs-less-than-10-cents/.

9   Kweilin Ellingrud et al., "Generative AI and the Future of Work in America," McKinsey Global Institute, July 26, 2023, https://www.mckinsey.com/mgi/our-research/generative-ai-and-the-future-of-work-in-america.

10 ARK Invest, "Big Ideas 2019," ARK Investment Management LLC, 2019, https://research.ark-invest.com/hubfs/1_Download_Files_ARK-Invest/White_Papers/Big-Ideas-2019-ARKInvest.pdf.

11 "Bob Lutz Predicts Non-Driverless Cars Will Be Illegal in 15 Years," *NESN*, November 8, 2017, https://nesn.com/2017/11/bob-lutz-predicts-non-driverless-cars-will-be-illegal-in-15-years/.

12 Neuralink, "A Year of Telepathy," *Neuralink Blog*, February 2025, https://neuralink.com/blog/a-year-of-telepathy/.

13 Eric A. Pierce et al., "Gene-Editing for CEP290-Associated Retinal Degeneration," *New England Journal of Medicine* 390, no. 19 (2024): 1827–1836, https://doi.org/10.1056/NEJMoa2309915.

14 Haydar Frangoul et al., "CRISPR-Cas9 Gene Editing for Sickle Cell Disease and ☒-Thalassemia," *New England Journal of Medicine* 384, no. 3 (2021): 252–260, https://doi.org/10.1056/NEJMoa2031054.

15 National Aeronautics and Space Administration, *NASA Climate Spiral Visualization*, visualization by Mark SubbaRao and Ed Hawkins, technical support by Laurence Schuler and Ian Jones, web administration by Ella Kaplan, science support by Gavin A. Schmidt, released November 15, 2023, https://svs.gsfc.nasa.gov/5190/.

16 Laure Resplandy et al., "Even if Emissions Stop, Carbon Dioxide Could Warm Earth for Centuries," *Princeton University News*, November 16, 2023, https://cmi.princeton.edu/news/even-if-emissions-stop-carbon-dioxide-could-warm-earth-for-centuries/.

17 The Royal Society, "Climate Change: Evidence and Causes," 2020, https://royalsociety.org/-/media/policy/projects/climate-evidence-causes/climate-change-evidence-causes.pdf.

18 NASA. "Mitigation and Adaptation." Last modified 2014. https://science.nasa.gov/climate-change/adaptation-mitigation/.

19 Susanna Twidale, "Climeworks Opens World's Largest Plant to Extract CO_2 from Air in Iceland," *Reuters*, May 8, 2024, https://www.reuters.com/business/environment/climeworks-opens-worlds-largest-plant-extract-co2-air-iceland-2024-05-08/.

20 Roman Krznaric, *The Good Ancestor: A Radical Prescription for Long-Term Thinking* (New York: The Experiment, 2020).

21 United Nations Environment Programme. "How Chernobyl Has Become an Unexpected Haven for Wildlife." Last modified September 16, 2020. https://www.unep.org/news-and-stories/story/how-chernobyl-has-become-unexpected-haven-wildlife.

22 "World's Largest 3D-Printed Neighborhood Built Using Special Robot—and Scientists Hope to Use the Tech on the Moon." *The Sun*, August 15, 2024. https://www.thesun.co.uk/tech/29771922/worlds-largest-3d-printed-neighborhood-built-robot-tech-moon/.

23 "Harmful Effects of the Microplastic Pollution on Animal Health." *International Journal of Environmental Research and Public Health* 19, no. 6 (2022): 3435. https://www.ncbi.nlm.nih.gov/pmc/articles/PMC9205308/.

24 Carl Sagan, *The Varieties of Scientific Experience: A Personal View of the Search for God* (New York: Penguin Press, 2006), 11.

25 Global Footprint Network. "Ecological Footprint Calculator." Last modified 2023. https://www.footprintnetwork.org/resources/footprint-calculator/.

26 World Commission on Environment and Development, *Our Common Future* (Oxford: Oxford University Press, 1987).

27 United Nations, *Transforming Our World: The 2030 Agenda for Sustainable Development*, A/RES/70/1, 21 October 2015, https://sdgs.un.org/2030agenda.

28 World Commission on Environment and Development, *Our Common Future* (Oxford: Oxford University Press, 1987).

29   "Could AI Create a One-Person Unicorn? Sam Altman Thinks So—and That's a Problem." *Fortune*, February 4, 2024. https://fortune.com/2024/02/04/sam-altman-one-person-unicorn-silicon-valley-founder-myth/.

30   António Guterres, "Secretary-General's Remarks to High-Level Opening of COP27," United Nations, November 7, 2022, https://www.un.org/sg/en/content/sg/speeches/2022-11-07/secretary-generals-remarks-high-level-opening-of-cop27.

31   Jack Flynn, "18 Average Screen Time Statistics [2023]: How Much Screen Time Is Too Much?" *Zippia*, March 10, 2023, https://www.zippia.com/advice/average-screen-time-statistics/.

32   Quixy Editorial Team, "25 Incredible Meeting Statistics: Virtual, Zoom & Productivity," *Quixy*, January 7, 2025, https://quixy.com/blog/meeting-statistics-virtual-zoom/.

33   "Telemedicine Use in the U.S. 2015-2022, by Channel," *Statista*, accessed February 25, 2025, https://www.statista.com/statistics/1219721/telemedicine-use-in-the-us-by-channel/.

34   "Wearable Technology Statistics," *TechReport*, https://techreport.com/statistics/wearable-technology-statistics.

35   Jeff Desjardins, "Charted: The World Has Passed 'Peak Child'," *Visual Capitalist*, January 21, 2025, https://www.visualcapitalist.com/charted-the-world-has-passed-peak-child/.

36   Renee Stepler, "World's Centenarian Population Projected to Grow Eightfold by 2050," *Pew Research Center*, April 21, 2016, https://www.pewresearch.org/short-reads/2016/04/21/worlds-centenarian-population-projected-to-grow-eightfold-by-2050/.

37   "ZERO Code: The Future Has Arrived." *Zero Energy Project*, May 14, 2018. https://zeroenergyproject.com/2018/05/14/zero-code-future-arrived/.

38   "How Did China Use More Cement Between 2011 and 2013 Than the US Used in the Entire 20th Century?" *The Independent*, March 25, 2015. https://www.independent.co.uk/news/world/asia/how-did-china-use-more-cement-between-2011-and-2013-than-the-us-used-in-the-entire-20th-century-10134079.html.

39   "Is Plastic Waste Really the New Asbestos?" *Safety and Management Solutions Ltd.*, accessed February 25, 2025, https://www.samsltd.co.uk/is-plastic-waste-really-the-new-asbestos/.

40   Charlton-Howard, Hayley S., Alexander L. Bond, Jack Rivers-Auty, and Jennifer L. Lavers. "'Plasticosis': Characterising Macro- and Microplastic-Associated Fibrosis in Seabird Tissues." *Journal of Hazardous Materials* 450 (2023): 131090. https://doi.org/10.1016/j.jhazmat.2023.131090.

41   Gordon E. Moore, "Cramming More Components onto Integrated Circuits," *Electronics*, April 19, 1965, https://www.cs.utexas.edu/~fussell/courses/cs352h/papers/moore.pdf.

42   Google AI Quantum and Collaborators. "Quantum Supremacy Using a Programmable Superconducting Processor." *Nature* 574 (2019): 505–510. https://doi.org/10.1038/s41586-019-1666-5.

43   Zhong, Han-Sen, Hui Wang, Yu-Hao Deng, Ming-Cheng Chen, Li-Chao Peng, Yao He, Jian Qin, et al. "Quantum Computational Advantage Using Photons." *Science* 370, no. 6523 (December 2020): 1460–1463. https://doi.org/10.1126/science.abe8770.

44   Hartmut Neven, "Meet Willow, Our State-of-the-Art Quantum Chip," *Google Blog*, December 9, 2024, https://blog.google/technology/research/google-willow-quantum-chip/.

45   International Human Genome Sequencing Consortium. "Initial Sequencing and Analysis of the Human Genome." *Nature* 409, no. 6822 (2001): 860–921. https://doi.org/10.1038/35057062.

46   Nidumolu, Ram, C.K. Prahalad, and M.R. Rangaswami. "Why Sustainability Is Now the Key Driver of Innovation." *Harvard Business Review*, September 2009.   https://hbr.org/2009/09/why-sustainability-is-now-the-key-driver-of-innovation.

47   National Institute of Building Sciences, *Natural Hazard Mitigation Saves: 2019 Report*, December 2019, https://www.nibs.org/files/pdfs/ms_v4_overview.pdf.

48   "Is HP Working on Eliminating Lead from Its Products?" *HP Sustainability*, last modified July 25, 2019, https://sustainability.ext.hp.com/en/support/solutions/articles/35000061794-is-hp-working-on-eliminating-lead-from-its-products-.

49   Peters, Tom. *The Little Big Things: 163 Ways to Pursue Excellence*. New York: Harper Business, 2010.

50   Gary E. Frank, "This Campaign Aims to Fight Climate Change With Cold Water Washing," *TriplePundit*, September 29, 2022, https://www.triplepundit.com/story/2022/climate-cold-water-washing/755806.

51   "European Recycling Platform (ERP)," ReTraCE, created 2002, https://www.retrace-itn.eu/partners/european-recycling-platform-erp/.

52   "Every Drop Counts at FedEx," 3BL Media, April 8, 2015, https://www.3blmedia.com/news/every-drop-counts-fedex.

53   "EPA Recognizes FedEx As Leader in Renewable Energy Use," *FedEx Newsroom*, January 29, 2008, https://newsroom.fedex.com/newsroom/united-states-english/epa-recognizes-fedex-as-leader-in-renewable-energy-use.

54   Amplify: "Amplify: Enterprise Strategy Execution Software." *Amplify*. https://www.amplify-now.com/.

55   Futureproof: "Futureproof: The Sustainability Platform for Businesses." *Futureproof*. https://poweredbyfutureproof.com/.

56   Persefoni: "Persefoni | Carbon Accounting, Decarbonization, and Climate Disclosure." Persefoni. https://www.persefoni.com/.

57   "Haystack Ag," Haystack Ag, accessed February 25, 2025, https://www.haystack-ag.com/.

58   "Voltpost Debuts Commercial Lamppost Electric Vehicle Charging Solution," Business Wire, April 11, 2024, https://www.businesswire.com/news/home/20240411301559/en/Voltpost-Debuts-Commercial-Lamppost-Electric-Vehicle-Charging-Solution.

59   "Sortile," United Nations Department of Economic and Social Affairs, https://sdgs.un.org/partnerships/sortile.

60   Martina Igini, "10 Concerning Fast Fashion Waste Statistics," *Earth.Org*, August 21, 2023, https://earth.org/statistics-about-fast-fashion-waste/.

61   "apic.ai," apic.ai, https://apic.ai/.

62   Margaret Lawrence, "Protecting Pollinators Critical to Food Production," *USDA National Institute of Food and Agriculture*, June 19, 2020, https://www.nifa.usda.gov/about-nifa/blogs/protecting-pollinators-critical-food-production.

63   "EarthScan," Mitiga Solutions, https://www.earth-scan.com/.

64   "Capabilities," Ellipsis Earth, https://www.ellipsis.earth/solution.

65   "Case Study: BCP Council, UK," Ellipsis Earth, https://www.ellipsis.earth/bcp.

66   Julia Finch, "Fast Food Firms Taken to Task After Survey of Street Litter," The Guardian, January 13, 2009, https://www.theguardian.com/business/2009/jan/13/fast-food-litter-mcdonalds-greggs.

67   Bradley, Omar N. "Set Your Course by the Stars." *The American Magazine*, March 1951, 31–33.

68   "Amundsen's South Pole Expedition," Wikipedia, last modified February 20, 2025, https://en.wikipedia.org/wiki/Amundsen%27s_South_Pole_expedition.
69   Carroll, Lewis. *Alice's Adventures in Wonderland.* London: Macmillan, 1865.
70   Leo Tolstoy, "Where Love Is, There God Is Also," trans. Nathan Haskell Dole (New York: Thomas Y. Crowell & Co., 1887).
71   Abraham H. Maslow, *The Farther Reaches of Human Nature* (New York: Viking Press, 1971).
72   Héctor García and Francesc Miralles, *Ikigai: The Japanese Secret to a Long and Happy Life* (New York: Penguin Books, 2016).
73   Albert Szent-Györgyi, *Bioenergetics* (New York: Academic Press, 1957)
74   Samuel Hulick, "Why People Don't Buy Products—They Buy Better Versions of Themselves," *Fast Company*, January 27, 2014, https://www.fastcompany.com/3025484/why-people-dont-buy-products-they-buy-better-versions-of-themselves/.
75   "InSight Captures a Martian Sunrise and Sunset," NASA Science, April 15, 2022, https://science.nasa.gov/resource/insight-captures-a-martian-sunrise-and-sunset/.
76   ALICE Collaboration, "ALICE Experiment," CERN, last modified August 2012, https://home.cern/science/experiments/alice.
77   "First Image of a Black Hole," NASA Science, April 10, 2019, https://science.nasa.gov/resource/first-image-of-a-black-hole/.
78   "Nokia: One Billion Customers—Can Anyone Catch the Cell Phone King?" *Forbes*, November 12, 2007.
79   "Best Inventions of 2007," *Time*, November 12, 2007.
80   Aryn Baker, "Zipline's Drones Are Delivering Blood to Hospitals in Rwanda," *Time*, October 13, 2016, https://time.com/rwanda-drones-zipline/.
81   "ICON and Lennar Announce Community of 3D-printed Homes is Now Underway in Georgetown, TX," ICON, November 10, 2022, https://www.iconbuild.com/newsroom/icon-and-lennar-announce-community-of-3d-printed-homes-is-now-underway-in-georgetown-tx.
82   Jim Euchner, "AI and Business Model Innovation," presented at the 2023 Innovators Summit, Innovation Research Interchange, October 3, 2023.
83   "Fraud Alerts and Notifications," American Express, https://www.americanexpress.com/en-us/credit-cards/credit-intel/fraud-alerts/.
84   David Wallace-Wells, "We Had the COVID-19 Vaccine the Whole Time," *New York Magazine*, December 7, 2020, https://nymag.com/intelligencer/2020/12/moderna-covid-19-vaccine-design.html.
85   "Moderna Announces Positive Interim Phase 1 Data for its mRNA Vaccine (mRNA-1273) Against Novel Coronavirus," Moderna, May 18, 2020, https://investors.modernatx.com/news/news-details/2020/Moderna-Announces-Positive-Interim-Phase-1-Data-for-its-mRNA-Vaccine-mRNA-1273-Against-Novel-Coronavirus/default.aspx.
86   Bernard Marr, "The Amazing Ways Babylon Health Is Using Artificial Intelligence To Make Healthcare Universally Accessible," *Forbes*, August 16, 2019, https://www.forbes.com/sites/bernardmarr/2019/08/16/the-amazing-ways-babylon-health-is-using-artificial-intelligence-to-make-healthcare-universally-accessible/.
87   "AI Compliance and Regulation Guide: What AI Companies Need to Know," *Koop.ai*, https://www.koop.ai/blog/ai-compliance-regulation-guide.
88   Grant Trahant, "Disrupting Healthcare Education: Osso VR's Impact on Surgical Training," *Causeartist*, November 20, 2024, https://www.causeartist.com/osso-vr-surgical-training/.

89   "Interactive Fashion Mirror | Virtual Fitting Room," VirtualOn, accessed February 25, 2025, https://virtualongroup.com/interactive-fashion-mirror-virtual-fitting-dressing-room/.

90   "IKEA Place app launched to help people virtually place furniture at home," IKEA, September 12, 2017, https://www.ikea.com/global/en/newsroom/innovation/ikea-launches-ikea-place-a-new-app-that-allows-people-to-virtually-place-furniture-in-their-home-170912/.

91   "Mixed Reality in Education," zSpace Blog, December 2024, https://blog.zspace.com/mixed-reality-in-education.

92   "Second Life is virtual world with real economy," Reuters, October 18, 2005, https://www.reuters.com/article/world/second-life-is-virtual-world-with-real-economy-idUSNOA630539/.

93   "Powerledger creates the world's first 'new energy' trading platform," Austrade, November 20, 2023, https://international.austrade.gov.au/en/news-and-analysis/success-stories/powerledger-creates-the-worlds-first-new-energy-trading-platform.

94   "Veridium to Use IBM Blockchain Technology to Create Social and Environmental Impact Tokens," *PR Newswire*, May 15, 2018, https://www.prnewswire.com/news-releases/veridium-to-use-ibm-blockchain-technology-to-create-social-and-environmental-impact-tokens-300648379.html.

95   "7 Benefits of IBM Food Trust," *IBM*, https://www.ibm.com/blockchain/resources/7-benefits-ibm-food-trust.

96   "Tracking the Future of Meat with Blockchain," Provenance, https://www.provenance.org/news-insights/tracking-future-meat-blockchain.

97   "Kalundborg Symbiosis: six decades of a circular approach to production," European Circular Economy Stakeholder Platform, https://circulareconomy.europa.eu/platform/en/good-practices/kalundborg-symbiosis-six-decades-circular-approach-production.

98   "Asnæs Power Station generates green power," Ørsted, November 2019, https://orsted.com/en/media/news/2019/11/asnaes-power-station-generates-green-power.

99   "HRH The Crown Prince inaugurated Asnæs Power Station," Ørsted, August 21, 2020, https://orsted.com/en/media/news/2020/08/217059569186945.

100   Hung-Suck Park et al., "Strategies for sustainable development of industrial park in Ulsan, South Korea--from spontaneous evolution to systematic expansion of industrial symbiosis," *Journal of Environmental Management* 87, no. 1 (April 2008): 1-13, https://pubmed.ncbi.nlm.nih.gov/17337322/.

101   "Shell to build one of Europe's biggest biofuels facilities," Shell, September 16, 2021, https://www.shell.com/news-and-insights/newsroom/news-and-media-releases/2021/shell-to-build-one-of-europes-biggest-biofuels-facilities.html.

102   Anne Marie Mohan, "Unilever innovates two new product formats for Loop," *Packaging World*, February 9, 2019, https://www.packworld.com/sustainable-packaging/article/13376897/unilever-innovates-two-new-product-formats-for-loop.

103   "P&G Joins TerraCycle's Loop – an Environmentally Friendly and Convenient E-Shopping Platform – With 11 Household Brands," Business Wire, January 24, 2019, https://www.businesswire.com/news/home/20190124005087/en/PG-Joins-TerraCycle%E2%80%99s-Loop-%E2%80%93-an-Environmentally-Friendly-and-Convenient-E-Shopping-Platform-%E2%80%93-With-11-Household-Brands.

104 Nestlé USA. (2019, January 24). *Nestlé joins TerraCycle as a founding partner of Loop, debuts reusable ice cream packaging.* Nestlé USA. https://www.nestleusa.com/media/pressreleases/nestle-joins-terracycle-founding-loop-reusable-ice-cream-haagen-dazs-packaging

105 PepsiCo. (2019, February 9). *PepsiCo elevates breakfast experience with high-end design aesthetics.* Packaging World. https://www.packworld.com/leaders-new/business-drivers-specialty/sustainability/article/13376949/pepsico-elevates-breakfast-experience-with-highend-design-aesthetics

106 Pauli, G. (2010). *The Blue Economy: 10 Years, 100 Innovations, 100 Million Jobs.* Paradigm Publications.

107 Katz, D. (2019). *Plastic Bank: Launching Social Plastic® revolution.* Field Actions Science Reports, Special Issue 19, 96–99. https://journals.openedition.org/factsreports/5478

108 Miriam Brusilovsky, "IDE Technologies | Desalination Can – and Does – Co-Exist in Harmony with the Environment," *International Desalination Association,* May 7, 2023, https://idadesal.org/ide-technologies-desalination-can-and-does-co-exist-in-harmony-with-the-environment/.

109 "Algae for Biofuel Production," *Farm Energy,* https://farm-energy.extension.org/algae-for-biofuel-production/.

110 "Ultrasound-Assisted Extraction of Microalgae: A Review," *PubMed Central,* accessed February 26, 2025, https://www.ncbi.nlm.nih.gov/pmc/articles/PMC9175141/. ² "Diatom," *Wikipedia,* https://en.wikipedia.org/wiki/Diatom.

111 "Ready Player DAO: The Game Has Changed," *Medium,* https://medium.com/@readyplayerdao/ready-player-dao-533c16dacf2f.

112 "Governance," *KlimaDAO,* https://www.klimadao.finance/governance.

113 Arun Sundararajan, "The Sharing Economy and the Evolution of Crowd-Based Capitalism," CFA Institute, August 9, 2017, https://blogs.cfainstitute.org/investor/2017/08/09/the-sharing-economy-and-the-evolution-of-crowd-based-capitalism/.

114 "Hello Tractor | Growing Together," Hello Tractor, https://hellotractor.com/.

115 "How Kiva Uses Slack to Crowdfund Loans for Underserved Communities," *Slack,* https://slack.com/customer-stories/kiva-slack-crowdfund-loans-underserved-communities.

116 "Why TransTech is the Future of Personal Transformation," Founder Institute, August 10, 2018, https://fi.co/insight/why-transtech-is-the-future-of-personal-transformation.

117 "Notes from the AI Frontier: Modeling the Impact of AI on the World Economy," McKinsey Global Institute, September 4, 2018, https://www.mckinsey.com/featured-insights/artificial-intelligence/notes-from-the-ai-frontier-modeling-the-impact-of-ai-on-the-world-economy.

118 European Parliament and Council of the European Union. 2016. "Regulation (EU) 2016/679 of the European Parliament and of the Council of 27 April 2016 on the Protection of Natural Persons with Regard to the Processing of Personal Data and on the Free Movement of Such Data (General Data Protection Regulation)." *Official Journal of the European Union* L119: 1–88. https://eur-lex.europa.eu/eli/reg/2016/679/oj.

119 California Consumer Privacy Act of 2018, Cal. Civ. Code § 1798.100 et seq. (2018).

120 United Nations. "Addressing Poverty." *United Nations Academic Impact.* https://www.un.org/en/academic-impact/addressing-poverty.

121 Prahalad, C.K., and Stuart L. Hart. "The Fortune at the Bottom of the Pyramid." *Strategy+Business,* no. 26 (2002): 54–67.

122 "Our Story," Toilet Board Coalition, https://www.toiletboard.org/about/.

123 "Consumer Unwillingness to Pay Extra for Sustainable Products: Impact on Small Businesses," Institute of Sustainability Studies, April 3, 2023, https://instituteof-sustainabilitystudies.com/insights/guides/consumer-unwillingness-to-pay-extra-for-sustainable-products/.

124 "Towards a Circular Economy: Fabrication and Characterization of Biodegradable Dinnerware from Sugarcane Bagasse," Frontiers in Sustainable Food Systems, accessed February 26, 2025, https://www.frontiersin.org/articles/10.3389/fsufs.2023.1220324/full.

125 Eliane Gluckman et al., "Hematopoietic Reconstitution in a Patient with Fanconi's Anemia by Means of Umbilical-Cord Blood from an HLA-Identical Sibling," New England Journal of Medicine 321, no. 17 (October 26, 1989): 1174–78, https://doi.org/10.1056/NEJM198910263211707.

126 Prisciandaro, M., Santinelli, E., Tomarchio, V., Tafuri, M. A., & Bonchi, C. (2024). Stem Cells Collection and Mobilization in Adult Autologous/Allogeneic Transplantation: Critical Points and Future Challenges. *Cells*, 13(7), 586. https://doi.org/10.3390/cells13070586

127 Ostrea, Enrique M., Jr., et al. "Meconium Analysis to Detect Fetal Exposure to Neurotoxicants." *Archives of Disease in Childhood* 91, no. 8 (2006): 628–629. https://www.ncbi.nlm.nih.gov/pmc/articles/PMC2083048/.

128 Laurie Boucke, Infant Potty Training: A Gentle and Primeval Method Adapted to Modern Living (Golden, CO: White-Boucke Publishing, 2000).

129 Ingrid Bauer, Diaper Free: The Gentle Wisdom of Natural Infant Hygiene (New York: Plume, 2001).

130 Yvette Z. Szabo and Danica C. Slavish, "Measuring Salivary Markers of Inflammation in Health Research: A Review of Methodological Considerations and Best Practices," *Psychoneuroendocrinology* 124 (December 1, 2020): 105069, https://doi.org/10.1016/j.psyneuen.2020.105069.

131 M. Gelardi et al., "Nasal Cytology: Practical Aspects and Clinical Relevance," *Clinical & Experimental Allergy* 46, no. 6 (June 2016): 785–92, https://doi.org/10.1111/cea.12730.

132 Elizabeth Best, "Beauty Review: FOREO's Luna 3 Cleansing Device Is A Skincare Revolution," *Embrace Brisbane*, published 4.4 years ago, https://embracebrisbane.com.au/beauty-review-foreos-luna-3-cleansing-device-is-a-skincare-revolution/.

133 "Kérastase and Withings Unveil World's First Smart Hairbrush at CES 2017," *L'Oréal Finance*, published 8.0 years ago, https://www.loreal-finance.com/system/files/publication-content/documents/LOREAL_kerastase_et_withings_2017_01_04_EN.pdf.

134 Francis Bacon, *Meditationes Sacrae* (London: Excusum impensis Humfredi Hooper, 1597).

135 "Menstrual Health & School Absenteeism in Laos: New Research Findings," Days for Girls International, accessed February 25, 2025, https://www.daysforgirls.org/blog/menstrual-health-school-absenteeism-in-laos-new-research-findings/.

136 Plan International UK. *Break the Barriers: Girls' Experiences of Menstruation in the UK*. London: Plan International UK, 2018. https://plan-uk.org/file/plan-uk-break-the-barriers-report-032018pdf/download?token=Fs-HYP3v.

137 Oni Lusk-Stover et al., "Globally, Periods Are Causing Girls to Be Absent from School," *World Bank Blogs*, June 27, 2016, https://blogs.worldbank.org/education/globally-periods-are-causing-girls-be-absent-school.

138   United Nations Department of Economic and Social Affairs. "Achieving Full Gender Equality Still Centuries Away, Warns UN in New Report." United Nations, September 7, 2022. https://www.un.org/en/desa/achieving-full-gender-equality-still-centuries-away-warns-un-new-report.

139   Janine M. Benyus, *Biomimicry: Innovation Inspired by Nature* (New York: Morrow, 1997).

140   George de Mestral, "An Idea That Stuck: How George de Mestral Invented the VELCRO® Fastener," Velcro Companies, November 2016, https://www.velcro.com/news-and-blog/2016/11/an-idea-that-stuck-how-george-de-mestral-invented-the-velcro-fastener/.

141   Frank E. Fish, "Hydrodynamic Design of the Humpback Whale Flipper," *Journal of Morphology* 225, no. 1 (1995): 51–60.

142   "Biomimicry Taxonomy," Biomimicry Institute, https://asknature.org/resource/biomimicry-taxonomy/.

143   Aristotle. *Physics*. Translated by R. P. Hardie and R. K. Gaye. In *The Complete Works of Aristotle*, edited by Jonathan Barnes, vol. 1, 315–446. Princeton: Princeton University Press, 1984.

144   Prelas, Mark, et al. "A Review of Nuclear Batteries." *Progress in Nuclear Energy* 75 (2014): 117–148.

145   Matthew W. Kanan and Daniel G. Nocera, "In Situ Formation of an Oxygen-Evolving Catalyst in Neutral Water Containing Phosphate and Co^{2+}," *Science* 321, no. 5892 (2008): 1072–1075, https://doi.org/10.1126/science.1162018.

146   Giacomo Ciamician, "The Photochemistry of the Future," Science 36, no. 926 (1912): 385–394, https://www.science.org/doi/10.1126/science.36.926.385.

147   Bennetto, H. P., G. M. Delaney, J. R. Mason, S. D. Roller, J. L. Stirling, and C. F. Thurston. "A Microbial Fuel Cell Capable of Converting Glucose to Electricity at High Rate and Efficiency." Biotechnology Letters 2, no. 4 (1983): 589-595. https://www.researchgate.net/publication/9043614_A_microbial_fuel_cell_capable_of_converting_glucose_to_electricity_at_high_rate_and_efficiency.

148   Laurent Lebreton et al., "Evidence that the Great Pacific Garbage Patch is rapidly accumulating plastic," *Scientific Reports* 8, no. 1 (2018): 4666, https://doi.org/10.1038/s41598-018-22939-w.

149   Leslie, Dick Vethaak, Marja H. J. R. Peijnenburg, Anthonie M. J. R. P. Kole, Jurriën A. J. G. van Velzen, and Heather A. M. Dick Vethaak. "Discovery and Quantification of Plastic Particle Pollution in Human Blood." Environment International 163 (2022): 107199. https://pubmed.ncbi.nlm.nih.gov/35367073/.

150   David F. Williams, "On the Nature of Biomaterials," Biomaterials 30, no. 30 (2009): 5897–5909, https://doi.org/10.1016/j.biomaterials.2009.07.027.

151   United States Department of Agriculture (USDA), Guidelines for Designating Biobased Products for Federal Procurement, Federal Register 80, no. 114 (June 15, 2015): 34023–34031, https://www.federalregister.gov/documents/2015/06/15/2015-14418/guidelines-for-designating-biobased-products-for-federal-procurement.

152   European Bioplastics, "What Are Bioplastics?" https://www.european-bioplastics.org/bioplastics/.

153   Kovatchev, Boris P., and Stacey L. Anderson. "Using Continuous Glucose Monitoring in Clinical Practice." *Frontiers in Endocrinology* 11 (2020): 1-10. https://doi.org/10.3389/fendo.2020.00149.

154   Aaron Long, "Remote Patient Monitoring for Chronic Disease Management—Care Beyond the Brick-and-Mortar," Decisio Health, February 6, 2025, https://decisiohealth.com/remote-patient-monitoring-for-chronic-disease-management-care-beyond-the-brick-and-mortar/.

155 "How Telemedicine Reduces Costs and Improves Patient Outcomes." Medesk. Accessed February 24, 2025. https://www.medesk.net/en/blog/telehealth-reduces-healthcare-costs/.

156 Luís Pinto-Coelho, "How Artificial Intelligence Is Shaping Medical Imaging Technology: A Survey of Innovations and Applications," *Bioengineering* 10, no. 12 (December 18, 2023): 1435, https://doi.org/10.3390/bioengineering10121435.

157 "More Breast Cancer Cases Found When AI Used in Screenings, Study Finds," *The Guardian*, January 7, 2025, https://www.theguardian.com/society/2025/jan/07/more-breast-cancer-cases-found-when-ai-used-in-screenings-study-finds.

158 "Artificial Intelligence in Medical Imaging," Spectral AI, August 16, 2024, https://www.spectral-ai.com/blog/artificial-intelligence-in-medical-imaging/.

159 "The Doctors Pioneering the Use of AI to Improve Outcomes for Patients," *Financial Times*, November 15, 2024, https://www.ft.com/content/2fd63023-ec0a-421c-9abb-b6c8000b3b51.

160 Bhagelu R. Achyut, Nadimpalli Ravi S. Varma, and Ali S. Arbab, "Application of Umbilical Cord Blood Derived Stem Cells in Diseases of the Nervous System," *Journal of Stem Cell Research & Therapy* 4 (May 7, 2014): 1000202, https://doi.org/10.4172/2157-7633.1000202.

161 Tayla R. Penny et al., "Umbilical Cord Blood Derived Cell Expansion: A Potential Neuroprotective Therapy," *Stem Cell Research & Therapy* 15, no. 1 (July 29, 2024): 234, https://doi.org/10.1186/s13287-024-03830-0.

162 Zubin Master, Kirstin R.W. Matthews, and Mohamed Abou-el-Enein, "Unproven Stem Cell Interventions: A Global Public Health Problem Requiring Global Deliberation," *Stem Cell Reports* 16, no. 6 (June 8, 2021): 1435–1445, https://doi.org/10.1016/j.stemcr.2021.05.004.

163 Francesco Dazzi and Mauro Krampera, "Mesenchymal Stem Cells and Autoimmune Diseases," *Best Practice & Research Clinical Haematology* 24, no. 1 (2011): 49–57, https://doi.org/10.1016/j.beha.2011.01.002.

164 Ray Kurzweil, *The Singularity Is Nearer* (New York: Viking, 2024).

165 Ray Kurzweil, "Humans Will Achieve Immortality by 2030," *New York Post*, March 29, 2023, https://nypost.com/2023/03/29/immortality-is-attainable-by-2030-google-scientist/.

166 Antonio Regalado, "Meet Altos Labs, Silicon Valley's Latest Wild Bet on Living Forever," *MIT Technology Review*, September 4, 2021, https://www.technologyreview.com/2021/09/04/1034364/altos-labs-silicon-valleys-jeff-bezos-milner-bet-living-forever/.

167 Eördögh, Fruzsina. «Russian Billionaire Dmitry Itskov Plans on Becoming Immortal by 2045." *Vice*, July 16, 2012.

168 Dmitry Itskov, "2045 Initiative," https://2045.com/.

169 Lencioni, Patrick. The Three Signs of a Miserable Job: A Fable for Managers (and Their Employees). San Francisco: Jossey-Bass, 2007.

170 Doran, George T. "There's a S.M.A.R.T. Way to Write Management's Goals and Objectives." *Management Review* 70, no. 11 (1981): 35–36.

171 Justin Weather. "Award-Winning Image of Earth Size Compared to Water and Air." Justin Weather, October 13, 2021. https://justinweather.com/2021/10/13/award-winning-image-of-earth-size-compared-to-water-and-air/.

www.ingramcontent.com/pod-product-compliance
Lightning Source LLC
Chambersburg PA
CBHW021759190326
41518CB00007B/368